Solar That Really Works!
Collyn Rivers
RVBooks.com.au (2020)

This virtually all-new edition explains, in clear English, every detail of designing and installing solar in boats, camper trailers, caravans and motorhomes. It is up-to-date, valid globally and technically accurate. Like all of Collyn Rivers' books it is written in plain English.

Publishing details

Publisher: RV Books, 2 Scotts Rd, Mitchells Island, NSW, 2430. info@rvbooks.com.au

Solar That Really Works!

Fifth Edition, 2020

Copyright: (2020) Collyn Rivers. All rights reserved. Apart from minor extracts for the purposes of review, no part of this publication may be reproduced, stored in a retrieval system other that of the buyer's own computer, or transmitted in any or form or by any means, electronic, mechanical, photocopying, recording or otherwise without the direct written permission of the publisher.

National Library of Australia - Cataloguing-in-Publication data

Rivers, Collyn

ISBN: 978-0-6483190-3-0.

1. Solar That Really Works 2.Solar in Motorhomes 3. Solar in Caravans 4. Recreational Vehicle Solar 5. Solar in Cabins 6. Solar in Camper Trailers. 7. Title.

Publisher's Note: To ensure topicality this book is updated as deemed required. The author genuinely appreciates reader and trade feedback relating to errors and omissions.

Disclaimer: Every effort has been made to ensure that the information in this publication is accurate however no responsibility is accepted by the publisher for any error or omission or for any loss, damage, or injury suffered by anyone relying on the information or advice contained in this publication, or from any other cause. The author has extensive experience in the design and construction of electrical equipment but is not a licenced electrician.

Chapter List

A complete table of content is available at the end of this book.

Preface	1
Chapter 1: Solar realities	3
Chapter 2: Electrical self-sufficiency	14
Chapter 3: Solar modules	18
Chapter 4: Solar regulators and monitors	26
Chapter 5: Batteries and battery charging (general)	31
Chapter 6: Batteries and battery charging (via alternators)	38
Chapter 7: Generators and fuel-cells	44
Chapter 8: Inverters	48
Chapter 9: Refrigerators	54
Chapter 10: Lighting	61
Chapter 11: Water and pumping	67
Chapter 12: Computers and TV	71
Chapter 13: Communications	73
Chapter 14: Scaling the system	79
Chapter 15: Example systems	89
Chapter 16: Extra-low voltage wiring	100
Chapter 17: Low voltage wiring	115
Chapter 18: Installing batteries	125
Chapter 19: Installing solar modules	132
Chapter 20: Installing the solar regulator	142
Chapter 21: Installing the fridge	147
Chapter 22: Installing an inverter	149
Chapter 23: Installing water systems	151
Chapter 24: Installing a voltage sensing relay	153
Chapter 25: Electrical converters	155
Chapter 26: Compliance issues	157
Chapter 27: Fixing problems	160
Chapter 28: Living with solar	163
Chapter 29: Walking the walk	167
Chapter 30: Electricity - simply explained	174
Chapter 31: Making contact	177
Detailed table of contents	181

Preface

Solar That Really Works! was initially available (in 2002) as two editions: one for caravans and another for motorhomes. Then, fifth-wheeler caravan and also camper trailer ownership rocketed. As these have characteristics of both caravans and motorhomes it made sense (in 2008) to combine the editions and also include cabins. A 'cabin' (in Australia and New Zealand) is classified in Electrical Standards as 'a structure designed for human occupation, being of one or more parts, and capable of being placed on or removed from a site.'

By 2011, changing technology necessitated a revised and expanded third edition. Growing sales enabled higher reproduction quality and more pages. A fourth edition was produced in August 2016, and an updated PDF version in June 2108. This version has again been updated and will be continue to be updated as/when required.

Powering a cabin or an RV lights and appliances from solar energy is neither difficult nor complicated. Nor is planning the system. Anyone comfortable with basic tools will have little difficulty installing and connecting the various (usually 12 volt) bits and pieces. Work on 230 volts, however, is legally required (in Australia) be done by a licensed electrician.

To make solar work successfully, we need to know how much solar energy is available where and when. We need to know what can realistically be run from solar energy and what cannot. We also need to know what to specify, buy and then do to make it all happen. Get this right and solar works superbly. Get it wrong and it doesn't.

This book is intended as a means to an end, not a text to be learnt. To avoid the need for memorising stuff that may never be needed again there is some deliberate repetition. For those with little or no electrical knowledge please initially read 'Electricity Explained' (Chapter 30).

Where feasible, this book uses everyday English. To avoid confusion, it uses technically correct (ISO) units and abbreviations and, for legal and other reasons, the legally prescribed term ('Extra.low voltage') for 12/24 volts where to do otherwise might confuse. 'Mains' voltage has been 230 V +10% to - 6% in Australia, New Zealand, UK and EU since 1995, and is known technically and legally (but sometimes confusingly) as 'Low-voltage'.

In the USA, the power supply (known there as grid power) is at a nominal 120 volts but 240 volts three-phase power is also widely used.

Commercial products named or mentioned are done totally without payment - and for the purposes of showing that typical. Endorsement is neither intended nor should be assumed.

The author discloses minor involvement (between 2006 and 2010) in the specifications and subsequent testing of the Redarc BMS 1215 unit, and also of long-term testing of Webasto diesel heaters. Both, however, were undertaken on a non-payment basis.

For those seeking to buy, design and build large solar systems for homes and properties the author also offers the companion volume Solar Success. Details are on solarbooks.com.au.

For in-depth coverage of vehicle electrics, our companion book Caravan & Motorhome Electrics too is written in substantially plain English. As with all our books, it sells globally.

Chapter 1
Solar realities

On clear days around noon, up to 1000 watts of solar energy (enough to boil a kettle in about five minutes), is theoretically available on each square metre of much of the Earth's surface. Commercially available solar modules (in 2020) convert only 20% or so of that energy into electricity. By using appropriate and efficient appliances, however, such solar can free recreational vehicles and cabins substantially or totally from mains, alternator or generator power.

Whether an RV, a cabin or even a big property system, the differences are mostly of scale. As apparently similar lights and appliances may use hugely different amounts to achieve much the results, a good starting point is to know approximately how much energy different things require.

If installed properly (and Chapter 21 shows how) today's up-market RV fridges draw only a third or so of the energy of many an eBay special. Microwave ovens draw far more current (i.e. electrical energy) than many people suspect. Done properly, pumping water needs little energy, done badly it needs a lot.

Where can solar energy be used?

It is light, not heat, that solar modules convert to electrical energy. The amount of energy they produce depends on how much light falls on them and for how long. They lose output when hot, so work best in cool places under a bright sun. All solar modules need at least some sunlight to operate. None work in total shade.

How is available solar measured?

A solar module's output is measured in so-called 'Peak Sun-Hours' (PSH). The PSH concept (conceived by the solar industry) is like using a rain gauge to measure a day's 'downfall. It is in general solar industry use but is not recognised or generally used academically.

Figure 1.1. Our previously owned OKA in Kakadu National Park (1998). Two 80 watt Solarex modules provided ample power for all needs - including a 71-litre Autofridge. During our ten-year ownership, the batteries did not once run out of power. The spade marks where the damper was cooking.
Pic: rvbooks.com.au.

The solar industry defines 1 PSH as an intensity of sunlight equivalent to that 'falling' on 1000 watts per horizontal square metre. That intensity varies with the sun's position in the sky and atmospheric conditions (such as haze and clouds). The peak input is usually at noon.

For the purposes of this book, each PSH can be seen is like a 'standardised drum full of sunlight'. That drum may 'fill' in only an hour or so in Australia's Cairns or Broome during most of the year, but may take all day during a Melbourne mid-winter. Each full drum can thus be seen as holding of 1 PSH.

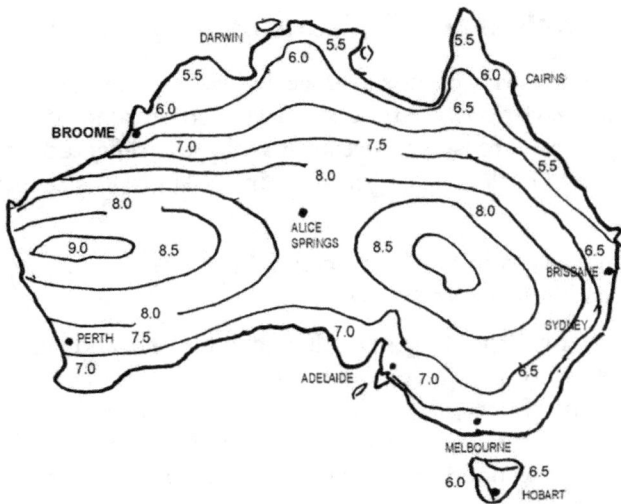

Figure 2.1. Peak Sun-Hours (mid-January). Multiplying the data shown by a solar module's true wattage gives the total average output for that day in watt hours/day. This map, plus that for mid-July, is reproduced at larger scale as Figure 1.13 in Chapter 13.

This book shows output in PSH. Multiplying the solar modules' true energy output (typically 70% of that seemingly claimed) by the PSH shown is the amount of the energy you can, on average, expect each day. This energy is measured (as we do at home) in watts.

The average amount of sunlight (irradiation) varies more or less linearly from mid-summer to mid.winter. Figure 2.1 shows PSH for a typical Australian January (mid-summer). As can be seen, in many places and times there will be at least 3 PSH each day; in some there will be 7 PSH or more. Full-size versions of these maps for Australian summer and winter are reproduced in Chapter 13 - as Figure 1.13. New Zealand's North Island, and the eastern part of the country's South Island have a fairly uniform 4.2-6.5 PSH between September and June, and 2-4 PSH in between.

Meteorological offices worldwide have solar irradiation maps for almost anywhere, but may need you to juggle scientific units.

Solar anomalies and limitations

Peak Sun-Hour maps allow for average seasonal cloud cover, but there are day to day variations. Output is usually high on sunny days that have light haze. It may increase yet further if sunlight is reflected from water or light sand and back to the haze layer, from where it is reflected down again.

Input typically halves during heavy cloud. Bush fire smoke may reduce it by two-thirds but it is rare to have no solar input.

Shortfalls resulting from long periods of cloud cover and night-time usage are covered by drawing on energy stored in battery banks. For most RVs the amount of energy storable is limited (due to battery size and weight), so larger systems, and particularly those with electric-only fridges, are likely to need generator or fuel-cell back-up (these issues are in Chapter 7).

The solar industry has an 'unusual' way of quoting this output that can cause buyers to expect 25%-30% more than they thought they had paid for. This particularly catches out those who understand electrics and/or physics - but are unaware of this 'marketing anomaly'. Chapter 3 explains all, but in essence you need to design the system assuming no more than 70% of the promoted solar module output. If ample space permits, 50% is safer.

Subject to the above, solar can be used successfully with RVs in most temperate areas between about 30 degrees north and south. By and large whatever works well in all parts of Australia (except Melbourne in mid-winter) is valid in most places where you are likely to use an RV. Differences in its scale and implementation, however, depend on the various needs, and (for most RVs) space and weight carrying ability. Solar capacity, however, is now so cheap that cost is rarely an issue.

Space and weight limitations

The load carrying capacity of a vehicle's axles, wheels and tyres is legislated - and also directly related to cost. Most caravan builders provide a so-called 'personal allowance' that rarely exceeds 250 kg and 350 kg respectively for single and twin-axle units. Included in that personal allowance' are gas, water, food and personal possessions: in essence everything placed in that caravan after it leaves the factory. 'Optional extras' (even if specified in the original contract) are usually installed by the dealer and thus likely to further reduce available allowance.

Campervans and motorhomes makers' major buyers are RV rental companies that seek to provide the maximum possible living space in vehicles still light enough to be driven by holders of a car licence. This has resulted in a caravan-like situation: ample loading space may be available but weight restrictions limit its use. This is less of an issue with larger specialist-built motorhomes: their load-carrying capacity is negotiable, and their length allows more space for solar modules.

Until recently, weight issues restricted battery capacity, but the much lighter lithium batteries now available substantially ease this. This is covered in Chapter 5.

Most fifth-wheel caravans (i.e. those that have their tow hitch above the tow vehicle's rear axle/s) have greater payload capacity (Figure 3.1). It is also usually feasible to house some part of the battery bank behind the cab of the towing vehicle, or in that vehicle's under-floor lockers.

Converter electrical systems

Almost all US made and now many locally-made RVs have 12 volt systems of which the battery back-up is intended only for occasional single overnight use away from 230 volt power. Unless extensively modified, these systems are close to useless for extended overnight camping. Chapter 25 addresses this.

Cabins

Cabins have fewer restrictions. There is usually ample space for solar modules and batteries. Theft was initially an issue but far less so since solar module cost dropped dramatically (post 2010).

Figure 3.1. This 11.3 metre fifth-wheeler built by Glenn Portch is exceptional in weighing only 3200 kg. It has a payload of an extraordinary 1300 kg!
Pic: Glenn Portch.

For cabins used irregularly, sealed lead-acid deep-cycle can safely be left permanently on charge as long as the necessarily high-quality solar regulator is programmed for the specific battery type.

The once popular 12 volt gel cell batteries are still made. Their main plus is that they are claimed to safely discharged to 10.8 volts, but cannot deliver high current.

Providing they are fully charged beforehand, AGM batteries may be left for 12 months or so before dropping below about a (non-damaging) 60% remaining charge at ambient temperatures below 25°C. It is generally best to avoid 'trickle charging' as the amount lost is so tiny that any form of long-term charging is liable to damage them (whilst otherwise rugged AGMs will not withstand overcharging).

Lithium iron batteries withstand even years of non-use without apparent damage, but the industry nevertheless suggests to initially leave them at about half charge for long term non-use.

Battery capacity

Until recently, solar capacity cost far more than battery capacity. Owners accordingly skimped on solar capacity: resulting in inadequately charged batteries that had their life limited accordingly. Furthermore, if the battery bank is overly-large relative to the charging source, that source may not be able to recharge it fully, let alone quickly. Adding more batteries alone is thus like opening more bank accounts for the same money deposited. It retains the same input as before -but increases your overhead losses.

Economise on batteries but never on solar modules. As a rough guide (for RV use) you need at least 200 watts of solar for every 100 amp hours of a 12 volt battery. Ideally have as much as the available space allows. There is no risk of overcharging as the associated solar regulator prevents this.

See also Chapter 29 re 'split systems' - where both tow vehicle and trailer each have their own self-contained (but interconnectable) systems. This is really worthwhile considering.

If you (improbably) carry energy-hungry arc welders and/or big angle grinders used only occasionally, scale the system for 'normal' loads and supply the rarely-used excess by a generator. This also applies if planning to spend only an odd winter month in places with short hours of sunlight (despite lower fridge energy usage in winter).

Cooking and heating

As roof space for solar modules is limited, solar generated electricity (alone) is not really practicable in RVs shorter than about 7 metres for anything that, as its main purpose, generates heat. Electric ovens, fryers, and water heaters are thus best avoided. Hair dryers are borderline. Electric irons are best used only where there is 230 volt mains power. For all but the largest RVs, use LP gas for cooking and heating water.

For cabins and the rare RVs that have ample space for solar modules and battery storage, it is feasible to use solar energy for cook tops (but less so for ovens). Use LP gas/solar water heaters for water heating generally.

Energy-efficient appliances

Coffee grinders, blenders and other small appliances vary in efficiency but, if used only occasionally, their energy use is rarely of concern. All microwave ovens, however, use more energy than many users suspect. Their wattage rating refers to the work they do (i.e. 'cooking power') not the energy used when doing so. Most '800 watt' rated microwave ovens consume about 1350 watts, or 1500 watts via an inverter. Ten minutes use may draw a day's output from a 100 watt solar module. That oven may thus cost only $195 or so, but running it from solar can add many times that for the extra solar capacity and battery capacity needed to drive it. It can still only be used when there's enough power. Excepting for big rigs with ample solar capacity, or a generator, consider running a microwave oven only when you have 230 volt mains access.

Water pumping

Apart from hand- or foot-operated pumps (both are still available), the only practicable pumps for RVs are those that run from 12 or 24 volts (Chapter 11). Mains-voltage pumps are available but they use far more energy for pumping the same amount of water.

Where there is a washing machine or dishwasher, and also in large cabins with flush toilets, a 'pressure accumulator' (Chapter 11) overcomes the otherwise high energy draw of pumping water. It also results in a system that does not fluctuate in pressure, is silent most of the time and saves electrical energy.

Washing machines/dishwashers

Most front-loading washing machines use less energy and water than top-loaders. The more efficient units run readily from a medium-sized RV solar system and inverter. They wash well using only cold water as long as cold water washing powder is used. These machines are fine also for cabins. Many current models draw only 200 watts or so when run from cold water.

Dishwashers need a hot water supply. It is not feasible to supply this (for RV use) via solar electricity, but a number of owners build their own thermal solar water heaters from coiled copper tubing or black poly pipe.

If doing so, to avoid scalding (especially of children) it is essential to include a 'tempering valve' (from plumbing suppliers) to ensure the water does not exceed 50°C. In some jurisdictions that valve is legally required.

Television

Figure 4.1. This Sony Bravia 32 inch TV draws 55 watts. Pic: Sony

Unless left on day-long, there are no major energy problems with recently-made TVs with 36 inch (92 cm) LED screens. Most draw 50-60 watts. Older ones may draw over 150 watts. Avoid any 12 volt TV made for sale in underdeveloped countries. They are ultra-cheap to buy but have energy-gobbling technology.

Computers

The larger laptop computers double as TVs, but as the screens are responsible for most of the draw their is no energy advantage. LED screens draw the least. See also Chapter 12.

Allow for the energy draw of charging iPads and also that drawn by communication modems. Be wary of specialised game-playing computers, they use far more energy, and tend to be used for longer.

Lighting

Incandescent (230 volt) globes are no longer legally sold in many countries. Halogen globes use about half the energy for the same amount of light but were an interim technology now mostly replaced by light emitting diodes (LEDs) - Chapter 10. Halogen globes were banned from sale in Australia in 2020.

Fluorescent globes and tubes and compact fluorescents use only a quarter of the energy of incandescent globes but the latest white and warm white LEDs use even less. Chapter 10 refers.

Air conditioning

Given at least 750 watts of solar modules for this alone, solar-powered air conditioning is feasible for daytime use, but unless backed up by mains electricity, or from a generator, having air conditioning all night is not practicable in any but the very largest RVs. Solar modules and air conditioners are, however becoming increasingly efficient. Later editions of this book may have a different view of feasibility.

Evaporative coolers use much the same energy as big cooling fans but lose effectiveness above 25% humidity. Their vendors often claim they work in up to 40% humidity. But vendors sell them - not necessarily use them.

What voltage?

Twelve and (the now rare) twenty-four volt systems are cheap and simple. Their wiring is relatively easy and for RVs and cabins in Australia and New Zealand) still legal to self-install. There is negligible risk from electric shock.

One drawback (for 12 volts) is that surprisingly heavy cable has to be used to reduce energy losses (Chapter 16). There is a wide range of 12 volt appliances, but (apart from fridges), very few for 24 volts.

Some coaches and a few motorhomes have 24 volt alternators and batteries. To run 12 volt lights and appliances, there are two ways of doing so.

Figure 5.1. Roof-mounted air conditioner. Pic: original source unknown.

Companies such as Redarc and GSL offer 24-12 volt charge equalising units. These draw the required 12 volts from one of the two series-connected 12 volt batteries used in most 24 volt systems, whilst constantly equalising the voltage across both batteries. A more efficient approach for lighter loads is to use a 24-12 volt dc-dc converter.

Mains-voltage via an inverter

An inverter provides mains-like electricity. Many seemingly identical units cost far less but may only be able to supply their rated maximum output for a second or two. They only seem identical. This is less of an issue with those over 1000 watts or so because they are made for a more electrically sophisticated market.

It is legal to self-install an inverter (made for this purpose) in an RV but only those into which appliances plug in directly (or via multi-outlet power board).

Those intended for connecting into a cabin or RVs fixed 230 volt wiring are of a different kind. They have no external socket outlets. They require specialised installation that (in Australia and New Zealand) must only be done by a licensed electrician. Chapter 17 explains more.

Costs

The cost of an RV or cabin's solar system is to some extent dictated by the fridge's system size and cost. With large ones, to safeguard against large-scale spoilage, it is advisable to have a back-up generator.

A microwave oven may cost only $195 but the solar capacity and battery capacity to drive it may add $1000. By all means have one, but (for use with small solar systems) run it only from a generator, or when you have access to mains power.

Solar is increasingly becoming cheaper, but battery capacity is not. Having adequate solar capacity alters the role of the battery. Sufficient battery capacity is still needed overnight and for dull days but, given sufficient solar, battery capacity can be reduced because solar modules charge to some extent even on overcast days.

Solar module price plummeted after 2010. Battery prices soared, but still differ considerably from vendor to vendor. It pays to compare prices but, because most batteries are so heavy, transport costs can wipe out otherwise seemingly bargain prices.

Small scale fuel-cell technology slowly continues. The initially promising Truma's VeGA LP gas.fuelled product (at 12,000 Euros) sadly proved far too costly. It was withdrawn from sale in 2014. The EFOY methanol-fuelled product is still on sale and rival LP gas-fuelled fuel cells are just (late 2019) appearing on the market. See Chapter 7.

Avoid cheap products

In the RV area particularly, unless you really know what you are doing, it is better to spend more and buy high-quality products from well-established companies rather than seeking bargain-priced products of unknown provenance and often negligible technical support. There are the odd bargains on eBay, but much is close to junk.

Chapter 2

Electrical self-sufficiency

To extend camping time away from mains power many RV owners add an auxiliary battery that is charged from the alternator whilst driving. It is a long-range fuel tank approach that works for single overnight stays, particularly for a fast 'around-Australia', where most days entail sufficient hours of driving to fully charge that battery. But for more typical usage, and much of the time, that battery is likely to be only partially charged and overly discharged. Both substantially shorten battery life.

A better general approach is where all energy (and energy loss) on site is replaced on site such that batteries primarily cater for overnight use, and supplement energy on overcast days. This, requiring solar, enables you to stay on site as long as you like (electrically at least). It costs more initially, but batteries last much longer. Savings on battery replacements partially compensate.

Doing this is feasible providing energy requirements are realistic and you are travelling where and when there is sufficient sun. This is most of Australia in summer and a fair part of it in winter.

It makes no sense however to scale a system for mid-winter usage in the colder less sunny areas, nor for occasionally running welders, big angle grinders, electrical clothes dryers. Such usage is best provided by a generator.

Vehicle alternator charging

Relying on all or part charging from the vehicle's alternator entails driving fair distances most days. It is done by parallel connecting the starter and auxiliary battery so that both charge from the alternator. Vehicles made prior to 2014 had a voltage sensing relay that delayed charging the auxiliary battery until the starter battery is adequately charged (typically inside a couple of minutes). As explained later in this chapter, this has become complex in recently-made vehicles. It is also possible that future emissions regulations may preclude using the vehicle alternator for this purpose - at least in fossil-fuelled vehicles. If so, the fuel cells described in Chapter 1 (and in depth in Chapter 7) will be the most probable alternative.

*Figure 1.2. Two 80 watt modules and AGM battery drove a 60 litre fridge in our previously-owned Nissan Patrol. At the time this picture was taken (2008), the system also included the very first Redarc BMS 1215 battery management system that we had on test for over two years. The TVan had its own 50 watt system for water pump, lights, NextG and laptop computer and a Webasto diesel water heater. Pic: En route to Mitchell Falls (Kimberley).
Pic: rvbooks.com.au*

Most alternators will charge a 100 amp hour battery in 2-3 hours but only to 70%-75% of full charge. AGM, gel-cell and lithium-iron (LiFePO4) batteries however charge more fully and faster.

There are products that overcome the voltage limitations of temperature-controlled alternators (Chapter 6) by directly increasing the voltage output of the alternator. Leaving aside the ethics of seeking to negate emissions reduction, and that increasing alternator voltage interferes with the vehicle's computer systems, such tampering with emission reduction is illegal.

The dc-dc units described later in this book do not cause the alternator to generate higher voltage. They accept whatever voltage that alternator produces and (via electronics) raises that output voltage within the dc-dc unit. In doing so this process reduces the current available. This is a minor compromise but works well.

The self-sufficient approach

Even right now, given adequate solar capacity, alternator charging is not essential. The author's rig shown in Figure 1.2. had (manually switched) provision for alternator charging, but such charging was never needed during six years of extensive use and ownership. Both tow vehicle and trailer had its own separate solar and battery system.

Omitting alternator charging, excepting for the starter battery, simplifies wiring and installation. There is also less to go wrong. Because the solar regulator can be set up specifically for the batteries used, this more readily enables a conventional battery to be used for starting and auxiliary AGM or lithium-ion batteries to be optimally charged for RV use.

Unless there is ample space for the solar modules required, this approach precludes fridges above 120 or so litres and also microwave ovens used for more than two or three minutes a day. The conventional solar plus alternator plus generator or fuel-cell approach (Chapter 6) is thus recommended.

The conventional solar plus alternator plus generator or fuel-cell approach (Chapter 6) is thus recommended for such use.

Jump starting

Figure 2.2. This (Wagan) jump starter will supply peak currents of up to 700 amps - more than enough to start the largest 4WD. Pic: Wagan.

Starting assistance can still be obtained from the auxiliary battery via jumper leads. A better way is to use one of the lithium battery (typically 18 amp hour) emergency engine starting units. Despite their small size and weight, these units will readily start a big 4WD engine several times before needing recharging. These units are often marketed as 18,000 milliamp hour. This is exactly the same thing as 18 amp hour but seems a great deal more if you do not understand electricity and/or the more curious ways of marketing.

Future for alternator charging

There is an ongoing trend to reduce the output of vehicle alternators to that required by the needs of that vehicle, plus minor allowance for (say) an upgraded sound system. It seems improbable that alternator output will be continue to be available for general RV use. If/when that happens, the fuel cells (described in Chapter 7) will be almost certainly be used instead.

Chapter 3
Solar modules

Figure 1.3. Monocrystalline solar module.
Pic: solarbooks.com.au

Solar modules convert the sun's radiation of light (between red and violet) into electricity. Ultra-violet light has too much energy (for use by present-day solar module technology) - it creates counter-productive heat.

The most commonly used solar modules have monocrystalline or polycrystalline cells. A further type uses a so-called amorphous technology that enables a solar module to be as thin as a human's hair and able to bend without damage, but at only 14% or so lower efficiency.

Monocrystalline modules

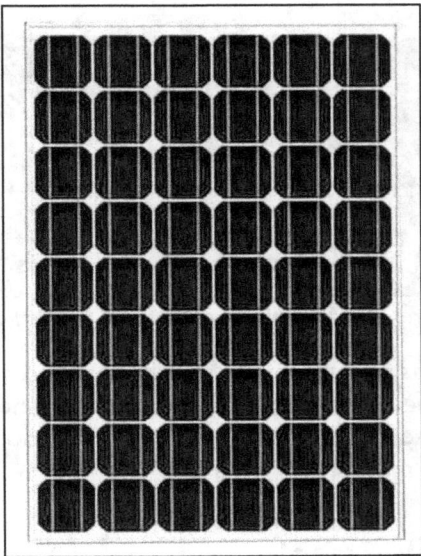

Figure 2.3. Polycrystalline solar module.
Pic: solarbooks.com.au

Monocrystalline modules have a uniform appearance. They usually cost more than polycrystalline but have higher efficiency, this is typically 16-20%, a few are about 20.5%.

As a generalisation, because they are usually smaller per watt, monocrystalline modules are preferable for RV use. They also produce more output under less light. Their life-span is about 25 years.

Polycrystalline modules

These modules are generally cheaper than monocrystalline and most are about 14%-16% efficient. A few makers, however, have developed polycrystalline modules that rival monocrystalline. As a generalisation, polycrystalline modules versions are better suited for cabin and residential use where mounting space is usually less of an issue.

Amorphous modules

Amorphous solar modules have better heat and shadow tolerance and were originally used where those qualities were important.

Figure 3.13. Amorphous solar panel. Pic: Redarc.

Their technology has so far limited their efficiency to about 14%. In the past few years, however, that they can be both thin and very flexible enabled them to be used more readily on curved RV roofs. They are also produced in roll-up mat form (Figure 3.3).

Solar cell development

By and large there have been only very small increases in solar cell output since 2016 or so.

The best (2019) solar cell technology still captures only 20% or so of the solar spectrum. SunPower's is 22.8%, LG's is 21.7% and Panasonic's is 20.3%. This is individual cell efficiency. As some 60 or more cells are soldered together to produce a (12 volt) solar module, that module's output will always be less than the best cell efficiency. Whilst close to 35% has been achieved in specialised applications it now seems that a totally different solar capturing technology is required for more general use.

What solar modules really produce

Unless used in conjunction with a so-called MPPT (Multiple Power Point Tracking) regulator (Chapter 4), solar modules produce only 70% or so of their apparently rated output. This is because the industry's SOC (Standard Operating Conditions), that are used also for promotion, are not typical operating conditions.

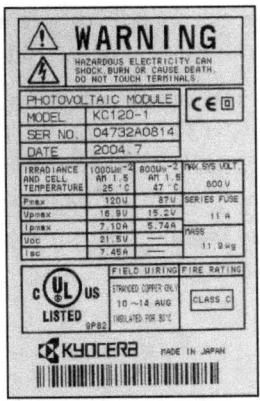

Figure 4.3. This data panel from one of the author's own solar system's modules marketed as 120 watts is shown as producing 87 watts under the NOCT rating.

All except amorphous solar modules lose about 5% of their output for every 10°C that they exceed a base-line of approximately 5°C. Vendors correctly claim that losses start at 25°C, but this misleads. That 25°C they refer to is not ambient temperature. It is that of little black solar cells under a hot sun. Those cells are typically 20°C warmer than ambient temperature.

Whilst not all vendors may reveal this (or are unaware of it) solar module makers do not conceal this. The industry's official NOCT (Nominal Operating Cell Temperature) method states that, at 25°C ambient, the actual cell temperature is typically +47°C. This can be seen in the third column of the photograph of an actual data panel (Figure 4.3). There, a '120 watt' solar module is shown as actually producing 87 watts.

There are further reasons why solar modules do not produce that seemingly claimed. People who understand electrics or physics are particularly likely to be misled - but in an unexpected way. They know that watts equal volts times amps so reasonably assume that solar industry wattage (for a solar module marketed as 12 volts) would be rated at that same voltage. But that's not how the industry does it.

To measure solar module output, the industry measures voltage and current separately. The claimed output is whatever combination (of volts and amps) results in the highest number of watts - and that's usually around 17 volts.

The industry justification is that if a nominally 12 volt load can withstand higher voltage without harm, and can thus do more work (as with many water pumps), a typically 16.8-17.2 or so volts will enable the load to reflect the claimed wattage output, (if you are also improbably on top of an equatorial mountain around noon on a cold but sunny day).

If, like most domestic users, you have a less improbable environment and a system that runs at 12-13 volts, you are likely to get only a bit over 70% of what you'd probably thought you'd paid for. A technique called Multiple Power Point Tracking (Chapter 4) can recover 10%-12.5% of this loss. The technology is now built into many solar regulators.

Shadow resistance

This term confuses many new to solar. No solar module works at all unless there is at least some light. Shadow resistance thus relates to how much output is lost when part of the module is shaded.

Basic solar modules lose almost all output if a third or more of the area has no sun. If half of such a module is shaded, there is next to no output. The latest and more advanced ones have inbuilt technology (including mini versions of Multiple Power Point Tracking mentioned above) that reduces but cannot eliminate such losses.

Module placement

Roof-mounted solar modules for RVs are best installed close to or totally horizontal. Apart from mid-winter the loss (across much of Australia is only a few percent).

For cabins, the modules should ideally face due north (in the southern hemisphere) and due south results in year-round optimum input. It is also feasible to have them variably tilted for (say) for a summer or winter holiday cottage for optimal seasonal input. The complex tracking mechanisms used in the early days of solar for home, cabin and occasional RV solar are now rarely used except in semi-arctic regions. This is mainly because solar module prices fell so dramatically after 2010 that, where space is available it became simpler and cheaper to add more capacity to compensate. Tracking is, however, now making a major comeback in large-scale solar farms.

Portable solar modules

A number of solar module and other vendors offer rigid solar modules hinged such that they fold together for storage. Hinged struts enable them to free-stand. They typically have a five or ten metre cable so you can have the solar module in the sun and your RV in the shade.

Figure 5.3. We experimented with a tilting solar module whilst crossing inland Australia in our rebuilt 1974 Kombi, but the gain did not warrant the effort. Author's wife (Maarit) is in foreground. Pic: rvbooks.com.au

The more basic units connect, via that cable, to the RV's existing solar regulator input. More costly versions have a solar regulator built into the solar module. This enables the portable modules to connect directly across the RV's battery - but the modules solar output will not be included in the RV's energy monitoring.

You can readily assemble your own portable solar modules using a pair of hinges and an adjustable strut for tilting them to face the sun.

Flexible versions of amorphous solar modules can be adhered to a curved RV roof (but some, using an aluminium backing) are reported as prematurely failing. Amorphous solar modules in 'roll-up blanket' form are also readily available.

Figure 6.3. High-quality roll-up solar modules. Pic: rolasolar.com.au.

Buying solar generally

When buying, decide what wattage and physical size/s best suit your needs.

According to the impartial pv-tech.org, the top ten solar module makers in 2018/9 are: Canadian Solar, First Solar (USA), GCL-SI (Hong Kong), JA Solar (China), Hanwha Q-Cells (South Korea), JinkoSolar (China), LONGi Solar (China), Risen Energy (China), Talesun (China) and Trina Solar (China).

Seek the lowest price for any of the above. Don't get talked into 'after-sales service'. Solar modules do not need any.

Almost all solar modules sold for RVs generate maximum power at 16.8-17.2 volts, then-reduced to the required safe charging voltage by the solar regulator.

Grid-connect solar modules for cabins and RVs

Solar modules made for grid-connect systems are often offered at very low prices. Most, however, develop voltages from 25-65 volts. They are not compatible with basic stand-alone systems that have nominally 12 volt solar regulators that typically accept up to 21 volts or so. These grid-connect modules are, however, often of high-quality. Most can be used for cabin and RV systems in conjunction with the MPPT solar regulators described in the following Chapter 4.

MPPT regulators

Multiple Power Point Tracking (MPPT) regulators, also described in detail in Chapter 4, accept a wide range of input voltages, yet can optimally charge 12 volt and 24 volt batteries. Minor expertise is needed to know if this is possible as the input voltage range of such regulators varies. Not all can be so used.

Solar module trends

Current solar module technology captures only part of the full solar spectrum. This limits the theoretically maximum achievable efficiency to about 35%. Close to this has been achieved in specialised applications, corresponding to about 350 watts per square metre, but such modules are hugely expensive space technology products.

It is now largely accepted that a totally different technology is required that enables a far wider 'light' spectrum to be exploited i.e. from near infra-red to far ultra-violet. It is unlikely however to expect their general commercial usage in less than a decade or so.

Chapter 4
Solar regulators and monitors

Solar regulators control the output from solar modules. They ensure the voltage input to the RV's lights and appliance will not exceed their safe upper limit and also that the RV's auxiliary battery/s charge quickly and deeply, but do not overcharge. Some owners still attempt to charge their batteries without a regulator, but this risks damaging both the batteries and connected equipment through excess voltage. Those claiming to get away with it mostly use a solar module that is too small to overcharge the battery, or have such grossly inadequate wiring that voltage drop 'safeguards' the system.

Figure 1.4. Basic Morningstar solar regulator is small but effective. Pic: Morningstar.

The simplest form of regulator is a basic switch that connects the module/s to the battery until it reaches full charge; the switch then opens and closes again when voltage drops. The more sophisticated regulators control charging current as well as charging voltage. Some turn on and off constantly at very high speed, automatically adjusting the ratio of on-time to off-time in each on/off cycle.

Most regulators are programmable for conventional batteries. This typically needs re-doing for each different battery type, voltage, capacity and time of day.

MPPT regulators

Figure 2.4. OutBack Power Systems solar regulator handles up to 70 amps. Pic: rvbooks.com.au.

Some part of the voltage mismatch loss (discussed in Chapter 3) can be partially regained by using a Multiple Power Point Tracking (MPPT) regulator. These devices juggle volts and amps to optimise the usable output (watts). When (say) a solar module is producing 17 volts at 5 amps, its theoretical and claimed output is 85 watts. But, as the *averaged* battery charge voltage is about 14 volts, most of solar output voltage from (say) 14 to 17 volts is not usable (so the actual wattage is 14 volts times 5 amps (70 watts).

An MPPT regulator works much like a torque converter in a car. It converts that input to about 14 volts at 5.5-5.6 amps - about 77-78 watts. The seemingly claimed 85 watts is still not there, but 77-78 watts is closer.

Some vendors claim that recovers 20%-30% of that 'lost'. Whilst true, that 20%-30% recovery occurs only for a few minutes early in the morning, and evening and/or when batteries are deeply discharged. A daily 10%-12.5% is more typical.

Some MPPT regulators accept solar input from 12-120 volts dc or more. Most are programmable to charge 12-48 volt batteries. This enables solar modules to be series-connected (Chapter 19), in turn enabling lighter interconnecting cable to be used. This can be a major saving with big solar arrays. MPPT regulators are also useful where a little more energy is needed, but there is no space available for extra modules.

Buying a solar regulator

Basic solar systems supplying an LED light or two plus a 12 volt water pump need only a basic $65-$70 regulator. But for anything more ambitious pay $200 upwards for a sophisticated solar regulator that includes monitoring functions.

Stand-alone monitors (as shown below) work well, and are simple to interpret, but some cost as much as good solar regulators that have monitoring functions inbuilt.

Most up-market solar regulators include MPPT technology and it is increasingly common in mid-priced units. Be wary, however, of claims of MPPT functionality for ultra-cheap eBay specials. Many are fakes (i.e. they not include MPPT).

Programming a solar regulator is reasonably easy once the manual has been read a few times. Teen-aged boys and girls are usually good at this, but rarely at explaining how.

Energy monitors - knowing the state of charge

Although solar regulators automatically control battery charging, you still need to know how much solar energy is available, how much is being used and how much energy remains in the battery. These functions are included within most solar regulators costing more than $200 or so, but often more clearly in stand-alone energy monitors, such as that shown in Figure 3.4. These can be located where easy to read.

Figure 3.4. This energy monitor is simple and very effective. Pic: Victron.

Do not attempt to assess deep-cycle battery charge by measuring its voltage. A conventional lead acid's battery's charge is the result of chemical interactions that take many hours before the charge becomes even throughout the battery. It is also temperature related (the colder the slower). Because of this any voltage reading during a charge or deep discharge and for a long time after, reflects only the local electro-chemical effect on the plates' surfaces and electrolyte close to them. Because of this, a well-charged such battery that has just run a microwave oven may show only 11 or so volts. This is often mistaken as being almost totally discharged. Conversely, a near-useless battery such battery may show 13-14 volts within minutes of being put on charge.

Instant voltage readings thus often result in totally wrong assumptions. As a virtually direct result, perfectly good batteries may be thrown away and those worn out retained. Meanwhile owners search pointlessly for 'faults' that may not exist.

AGM batteries respond faster, but still need 'resting' before any voltage measurement has meaning.

With the increasingly popular lithium iron batteries, their voltage in cabin and RV use stays almost constant (at 13.1-12.9 volts) across their recommended working range - of 90%-20% charge.

How energy monitoring works

For all rechargeable batteries the only truly practical way of knowing remaining charge works much as we track money. Count what comes in, count what goes out and deduct one from the other: the balance is what we have left.

Energy monitoring may be added to existing systems, but most of the functions offered are standard inclusions on good quality solar regulators.

The unit shown is an example of a good and popular stand-alone energy monitor. It is marketed under various names including Xantrex, Victron and Enerdrive. The functions typically include:

* Battery voltage.

* Lowest and highest battery voltage since mid-night.

* Current flowing in and out of the battery.

* Consumed amp hours.

* State of charge.

* Time to go (how long batteries can support an existing load).

* Historical data.

The need for battery monitoring may not be immediately obvious but it will quickly become so. Further, because most units record historical data, such monitoring eases fault finding.

Chapter 5

Batteries and battery charging (general)

This Chapter primarily relates to batteries being charged by other than vehicle alternators made later than year 2000. Post-2000 alternators and alternator charging is more complex (it is covered in Chapter 6).

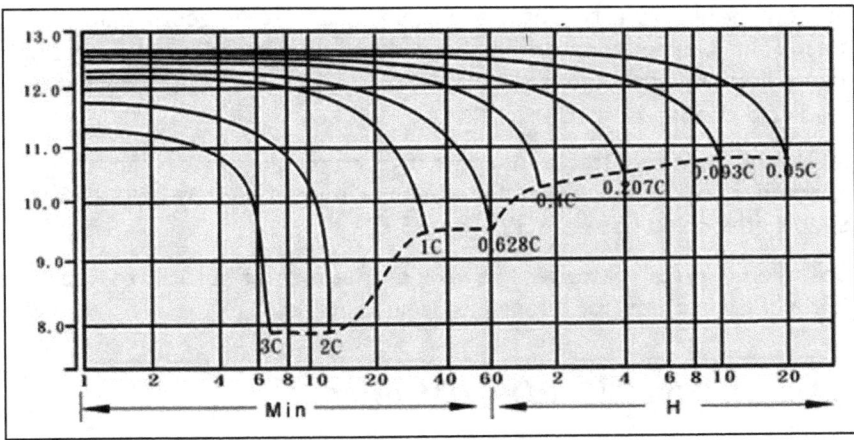

Figure 1.5. How battery available capacity relates to time current is drawn. Vertical scale is volts, horizontal is time. Graph: Long Chan Battery Co.

Traditional lead-acid batteries store energy as chemical reactions between lead plates and an acid solution. They are made in various types, sizes and applications. The available capacity drops rapidly as load current increases. This is often misunderstood to imply that remaining capacity is reduced. It is still there, but only accessible at a lower rate of discharge.

The available capacity drops rapidly as load current increases. This is often misunderstood to imply that remaining capacity is reduced. It is still there, but only accessible at a lower rate of discharge. A deep-cycle lead acid battery's life-span is primarily the number of amp hours in and out. This can be a small number of deep discharges, a high number of shallow discharges, or somewhere in between, but the relationship is not linear. Lead acid battery makers suggest discharging to 50% gives optimum life span and may enable 500-600 cycles of such use. Two thousand or more such cycles are not uncommon if you follow the self-sufficiency recommendations in this book, (such that batteries rarely drop below 75%-80% of full charge). Traditionally, correctly sized and maintained top grade deep-cycle batteries have outlasted all other types, but unless they are so used, so-called 'traction batteries' are a cheaper substitute. Ultimately however, how you use lead acid batteries is a monetary issue. It is almost as if you are buying usable amp hours.

Specialised batteries

Gel cells and AGM (Absorbed Glass Mat) batteries are larger, heavier and more costly than conventional lead acid batteries, but offset by greater usable capacity. Both have their electrolyte in gel or matrix form. This enables them to be charged and discharged faster. Ongoing deep discharging shortens their lives, but less so than with conventional lead acid deep-cycle batteries.

AGM and gel cell batteries need next to no maintenance. If first fully charged, they may be stored for a year or more if mostly below 25°C or six months or so between 30°C to 35°C.

So-called 'Crystal' batteries are now on the market but it is too early to know if their claim (for a long life) is sustainable.

Lithium-iron (LiFePO4) batteries

There are two main types of lithium batteries: lithium-ion and lithium iron. The former (not used in cabins or RVs) stores more energy but is less stable thermally.

Figure 2.5. Typical LiFePO4 batteries. Pic: Revolution Power.

Lithium-iron phosphate (often abbreviated as its chemical formula of LiFePO4) batteries are increasingly used in cabins and RVs. They are about a third the weight and volume of most other forms of rechargeable batteries. They charge much faster and can be discharged routinely to greater depth without overly shortening their typical life of about 2000 charge/discharge cycles (about 5.5 years if daily).

Nominally 12 volt LiFePO4 batteries in typical cabin and RV use maintain 13.1-12.9 volts across most of their usable range of about 90% to 20% charge. They require a specialised and critical way of charging. These batteries are made up of individual (nominally 3.4 volt) LiFePO4 cells that must be equally balanced, and charge voltage not allowed to exceed a critical level, nor to fully discharge.

Early users bought LiFePO4 cells that they interconnected for a nominal 12.8 volts, and added their own monitoring and control systems. For general RV acceptance however LiFePO4 batteries must be drop-in replacements for deep-cycle batteries and chargeable by that system existing. Some such are now available, but (as of 2020) they require specific charging regimes that are still included only in a limited range of high quality battery chargers.

LiFePO4's are excellent batteries and ideal for RV use, but must be correctly installed and charged.

High current loads - caution

If powered via an inverter, a typical 800 watt microwave oven draws 130 amps at 12 volts. Whilst lead acid deep-cycle batteries can safely supply current at a rate of 5% of their Ah capacity, their lives are thus substantially shortened if routinely used to drive a microwave oven. If running a microwave oven from battery power it is best by far to use LiFePO4 - or if not affordable, an AGM.

How batteries are charged

rechargeable batteries of all types are charged by applying a voltage across them that is higher than the existing voltage across them. The greater that charging voltage difference, the higher the charging current, and the faster the battery charges.

Earlier vehicle alternators and cheap mains battery chargers generate an approximately constant 14.2-14.4 volts but, as the charging battery's voltage rises, that voltage difference between battery and charger accordingly decreases.

Charging a battery from a constant voltage is cheap and simple, but is always a compromise. If the applied voltage is high enough to charge it in a realistic time, but then not cut off once fully charged, that voltage will damage the battery. Some chargers apply a voltage such batteries reach about 70% (of full charge) in eight to ten hours, but may take a day or two to fully charge. If accidently left on charge for a week or more the battery is likely to be damaged or destroyed.

Starter batteries nowadays are rarely a problem. Whilst starting a cold 4WD requires up to 500 amps at 12 volts, that draw is only for two to three seconds. It depletes the starter battery by a negligible 2% or so, and is replaced by the alternator within two to three minutes.

AGM and gel cell batteries accept higher rates of charge. They charge fairly well from constant voltage, and better and faster via the regime outlined below (providing the charging source has enough capacity). This regime, with minor variants, is built into virtually all high-quality chargers. LiFePO4 battery charging is discussed later in this chapter.

Typical battery charging sequence

Charging lead acid, gel and AGM batteries is usually done in three stages.

During the first (boost) stage, as battery voltage rises, so does charging voltage. This constant voltage difference is automatically adjusted to ensure that, within the maximum current capacity of the charger, the charge current stays at the highest rate that is safe for both the battery and for the charger itself. If the charger is large enough relative to the battery, that highest safe rate is typically 15%-25% of the battery's amp hour capacity. This stage typically brings the battery to 70%-75% of full charge.

During the second stage (absorption), the charging voltage (and hence charging current) is typically reduced such that the battery charges at a constant 10%-15% of its amp hour capacity. This, typically for two to three hours, enables the charge to become evenly distributed throughout the plates and electrolyte. If a heavy load is applied during this period, most such chargers revert to their boost cycle.

Following absorption, charging is reduced such that it counterbalances battery internal losses and minor constant loads (e.g. electric clocks, security alarms). This so-called 'floating' stage is at a voltage dependent on battery type and (ideally) temperature. It is typically 13.2-13.8 volts for conventional batteries and 13.2-13.6 volts for AGM and gel cell batteries (the lower voltages typically apply to high temperatures).

At this stage, a battery is likely to be 95%-100% charged. As long as the float voltage is correct and water level (if still applicable) checked from time to time, conventional lead acid batteries may be left floating indefinitely, but most AGM makers prefer their products to be differently treated.

AGM batteries have minimal internal leakage enabling them to retain about 60% charge for a year or more in temperate climates. Because of this, even minor float charging (for either type) may be overcharging. Some of the makers suggest that users planning to lay up their vehicles should fully charge the batteries before storage, and recharge them at about six month intervals in hot climates, and every twelve months where it is mostly below 25°C.

Equalisation

Now rarely used, this (usually manually-selectable cycle) pumps current through the fully-charged battery at up to 16 volts for an hour or two. Its intent is to overcharge to ensure all cells have equal voltage.

Whilst possibly desirable for the multiple series-connected two volt cells in stand-alone property systems, most battery makers say it is neither necessary nor desirable with today's battery technology.

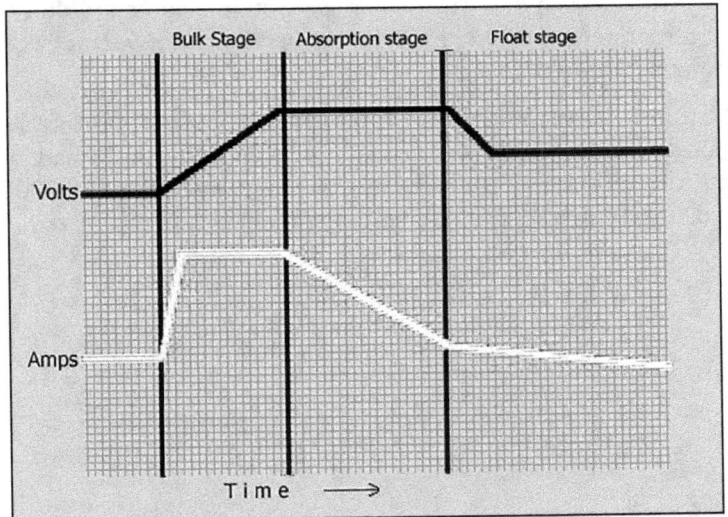

Figure 3.5. The basics of multi-stage (e.g. dc-dc) charging. Different products have minor variations but the general principles and intent are similar.

Do not use this (usually optional) cycle unless the battery maker specifically recommends it. Never equalise-charge an AGM battery. It is likely to destroy it.

Battery Management Systems

Figure 4.5. Early BMS 1215 undergoing off-road test. Pic: rvbooks.com.au.

In year 2000 or so, a number of companies began to develop and market systems that using basically the above charging sequence, combined all the required parts into one unit. A typical such unit may thus have a mains voltage battery charger, an alternator charger (with provision for solar input), a 230 volt inverter and full energy monitoring.

These systems originally met with buyer concern, that failure of one function could require the whole unit to be replaced. Buyers were, however, overlooking that most separate system failures are due to faulty interconnections and non-total compatibility between devices from different makers.

That best known was the Redarc BMS 1215. To check its reliability the first (pre-production) unit was bolted rigidly to a steel bulkhead in (the author's fully off-road OKA - Figure 4.5). It survived over 170,000 km of mainly dirt road driving over ten years and still works to this day.

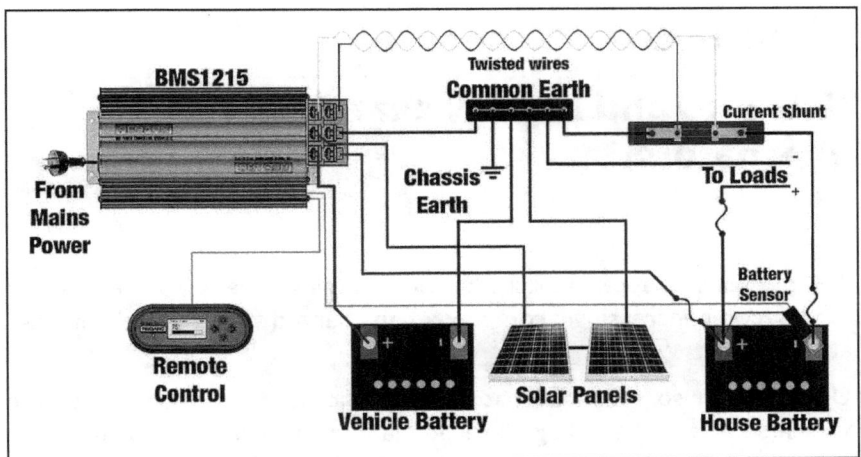

Figure 5.5. How a typical battery management system is installed (here a Redarc BMS 1215). Most interconnections are made within the unit itself. Drawing: redarc.com

These BMS systems are now made by many companies. They are by far the best way to install solar and alternator charging in RVs. The basic standard units work well with most vehicle alternators made prior to year 2014 or so. From that year on, as Chapter 6 explains, many vehicles have variable voltage alternators that require BMS systems designed for those units.

Chapter 6

Batteries and battery charging (via alternators)

The previous Chapter explains battery charging generally. It has however become increasingly complex when charged via a vehicle's alternator.

Until 2000 or so, most alternators had sufficient output to maintain the vehicle's battery at a voltage level adequate to run whatever needed in normal driving whilst also charging an RV's auxiliary battery.

To ensure adequate RV auxiliary battery charging, units were developed that electrically isolated the auxiliary system from the alternator, and boosted voltage to that optimally required for the RV battery. These units, now known generically as dc-dc chargers, are perceived by the alternator as just another load (e.g. as spotlights). Also installed was a voltage sensing relay, that opened if starter battery voltage dropped below about 12.6 volts to ensure the RV's auxiliary energy demand did not discharge the starter battery.

Temperature controlled alternators

Around year 2000 the EU's emissions regulations were tightened to reduce the engine warm-up period and to further reduce permitted carbon monoxide and diesel particulate limits. Whilst these requirements did not directly address vehicle electrics, car makers realised that limiting unnecessary alternator output would reduce the energy otherwise required. This did not make a major reduction in emissions, but enough to be worthwhile.

The result was alternators that produced about 14.2 volts when cold, and dropped to about 13.6 volts after the engine reached normal running temperature. Here too, the now increasingly accepted dc-dc chargers substantially overcame auxiliary battery charging issues.

Variable voltage alternators

Emission regulations tightened yet further in 2014. This resulted in many car makers changing to variable voltage alternators. These charge over a wide range of voltage - from well over 15 volts to 12.3 volts (some to zero).

A further change, primarily in electric and hybrid vehicles is to utilise the energy (known as kinetic energy) normally lost (as heat) whilst braking, and prolonged deceleration. There are various ways of doing so including storing the recovered energy in a flywheel, battery, hydraulics, and ultra-capacitors. The hydraulic system is being adopted in the utility vehicle segment. Hydraulics have more power density, better potential efficiency, and better energy storage efficiency, which augments its growth to only 80% or so. The vehicle's computer system detects any braking and increases the alternator's voltage and current (to 15 volts plus and at up to 200 amps). This slows the vehicle and the recovered 'braking energy' force-charges the 'main battery', boosting it to 100%. Alternator voltage is then reduced to about 12.3 volts, or the alternator is shut down altogether until the battery charge has dropped again to 80%.

In another (electrical) system the starter battery (now called the 'main' battery) is normally charged to only 80% or so. The vehicle's computer system detects any braking and increases the alternator's voltage and current (to 15 volts plus and at up to 200 amps). This slows the vehicle and the recovered 'braking energy' force-charges the 'main battery', boosting it to 100%. Alternator voltage is then reduced to about 12.3 volts, or the alternator is shut down altogether until the battery charge has dropped again to 80%.

The vehicle's main battery is designed specifically for this purpose and withstands the high voltage and currents involved. An over 15 volt charge (at an available up to 200 amps), however, wrecks lead acid deep-cycle, AGM and LiFePO4 batteries. This precludes directly connecting them across the 'main' battery whilst charging.

A further alternator issue (for RV use) is not just that plus 15 volt charge, but that charging frequently drops to 12.3 volts (and, with some alternators, to zero volts). This negates using a voltage sensitive relay (alone) to isolate the auxiliary battery. Were one to be used with such an alternator, whenever the output voltage drops below about 12.7 or so volts (which is much of the driving time) that relay opens for a typical two-three minutes, totally precluding auxiliary battery charging altogether each time it occurs.

These alternators are fitted to many new vehicles from 2013 onward. Most post-2017 vehicles have such alternators, and some have regenerative braking.

Figure 1.6. Typical Euro 5 alternator - note the large pulley size. Pic: original source unknown.

Identifying alternator type

If the alternator type is not revealed in the vehicle's specifications, it can be established by monitoring the voltage across the starter battery via a voltmeter or multimeter. Secure the leads as they otherwise tend to get caught up in the fan belt. Then drive at varied temperatures, conditions and loads (such as the air conditioner). Make a note of the lowest voltage encountered. If ever below 12.7 volts that alternator is odds-on to be variable voltage. See also redarc.com.au/calculator/ dual-battery-calculator.

Contrary to the advise given by a well known UK company, that minimum voltage cannot reliably be measured at the vehicle's cigarette lighter outlet. This is because many vehicles with variable voltage alternators now have a stabilised power supply that ensures that cigarette lighter output is maintained at a constant 12.8 volts.

Battery-to-battery DC chargers

To cope with variable voltage alternator issues (for RVs) companies such as Redarc (Australia) and Sterling Electrics (UK) have further developed so-called Battery Management Systems (Chapter 5).

Known generically as Battery-to-Battery DC chargers, these units use the vehicle's 'main' battery voltage to determine when and how to charge the RV's auxiliary battery, and when to separate the 'main' battery from the auxiliary battery to safeguard engine starting.

As with dc-dc charging, battery-to-battery dc charging accepts the variable alternator voltage output and boosts or reduces it to maintain a stable output using the charging profile that best suits the capacity and type of auxiliary batteries. It does so, as with most upmarket mains battery chargers, in the generally similar manner described in Chapter 5.

These systems work well with traditional lead acid batteries and AGMs, but not necessarily with lithium iron batteries.

Charging lithium-iron batteries

LiFePO4 charging should only be done via a charger that is known to be suitable for this purpose or has a specific LiFePO4 charging mode. There is still some lack of agreement between users and installers, regarding the depth of charging. Overcharging destroys or badly damages these batteries and is only a tiny voltage difference between full charge and overcharge. All require the cell management system mentioned in Chapter 5.

Whilst RV Books recommends the use of LiFePO4 batteries, it is essential to buy them, plus whatever required for charging them, by a company that truly knows how - or unless you really do know what you are doing.

Choosing a battery-to-battery charger

To suit various types of alternators, Redarc (for example) has a standard range of such chargers for vehicles where alternator voltage never falls below 12.7 volts whilst driving. The company's LV (Low Voltage) variants cope with variable voltage alternators and provide specific battery charging algorithms to suit lead acid, gel, AGM and calcium batteries. Some also accept solar input.

Figure 2.6. Redarc BCDC 1240-LV will handle 40 amps from ECU controlled variable voltage alternators. Pic: redarc.com

These and similar units from other makers overcome the limitations of varying alternator voltages, and fixed-voltage charging and are really the only charging methods (for RV batteries) now worth considering. All accept whatever alternator and solar voltage is available and increase or decrease it, using sequences generally similar to that shown by Figure 3.5) in Chapter 5. Some include a multi-stage 230 volt charger and/or accept solar module input.

Adding solar

Many people wish to run a large electric fridge or fridge-freezer (see Chapter 9). This is feasible via solar for big motorhomes, fifth-wheel caravans and coaches that have sufficient space for the substantial solar array required, but back-up alternator, fuel-cell or other auxiliary charging is advisable. An alternative is to use a T-rated fridge (Chapter 9) that runs from solar and the alternator whilst driving, 230 volts when available and LP gas at all other times. Doing so reduces the needed amount of solar and battery capacity.

By and large, there are no major problems if you wish to use both alternator and solar charging simultaneously and for several hours each day. It does not work as well as might be expected because, beyond 50% charge, the battery charges mainly from whichever source provides the higher voltage at any time. As the draw from one source goes up, the other goes down.

Whilst people tend to worry about doing so, it is fine to connect the output from a solar regulator directly across the RV's auxiliary battery despite it being also alternator charged.

Need for buying caution

Many low-end 230 volt battery chargers are sold as multi-stage chargers, but are not. They charge at about 14.4 volts and should turn off when the battery reaches that voltage. Not all do. Many a costly battery has its life shortened by this. Good chargers are costly, but this is not an area for skimping.

A good charger is programmable for the batteries you intend to charge, e.g. conventional lead acid, sealed lead acid, gel cell, AGMs etc. A good three-stage 15-20 amp such charger will outperform 30-40 amp constant voltage chargers once beyond 40% or so charge.

If planning to be extensively away from mains power, stand-alone solar necessitates daily solar input to be, on average, at least 15% more than that used, preferably 25%. Later sections of this book show how to assess the feasibility.

Chapter 7
Generators and fuel-cells

The most general need for a generator in an RV or cabin's solar electric system is to run an electric refrigerator when battery energy is low. Even if you have a coach-roof full of solar modules, a generator is usually necessary if you need air conditioning all night.

It is also advisable to use a generator for large infrequent loads such as clothes dryers (if run on a 'hot' cycle), and large power tools, or to limit their usage to where there is mains power. There is also a case for generator charging for RVs that spend only the odd winter month or two in cold areas. It is not worth increasing the solar and battery capacity otherwise not needed for the rest of the year.

Despite the Australian government's Product Emissions Standards Act 2017, that initially covered outdoor power equipment (under 19 kW) such as lawn mowers, generators, pumps and chain saws little, until recently most appear to have been all but totally ignored.

Figure 1.7. This Honda inverter generator produces up to a constant 2700 watts. Noise level is 58 dBA at 7 metres at its rated load.
Pic: Honda.

The Product Emissions Standards Rules was amended to allow the sale of remaining uncertified products before 1 July 2019, but truly basic generators can still be bought for only $100 or so. They are hideously noisy and polluting and have a very real risk of damaging computers, TVs etc connected to them.

The supply of products will become a legal offence on 1 July 2020.

This restriction is timely as independent research (e.g. that from Blue Sky Alliance) revealed that some such generators emitted up to 40 times that of a 1990s car. Similar regulations are also now being introduced in India and many other countries.

Safeguarding electrics

Always switch off electrical equipment before a basic generator runs out of fuel. When this happens the generator tends to 'splutter' to a halt. The resultant rapid speed-changes generate high voltage spikes that may damage the generator and whatever is connected to it at the time. This problem does not affect the inverter/generators described below.

Inverter generators (such as that shown in Figures

1.6. and 2.6) cost far more than the basic units described above. Nevertheless, these are the only types to consider for cabin and RV use as they are quiet and produce 230 volts ac that is cleaner than many a grid 230 volt supply.

Charging from an inverter generator

Most such units also provide 13.2 - 13.6 volts dc at an industry-common eight amps. This output is intended for directly powering 12 volt devices (i.e. where no battery is involved). Whilst that outlet may also be labelled 'battery charger', that voltage is too low bring a battery much beyond 50% charge. This is rarely disclosed by vendors. As a result generators plug away all day long in costly and pointless attempts to increase charge.

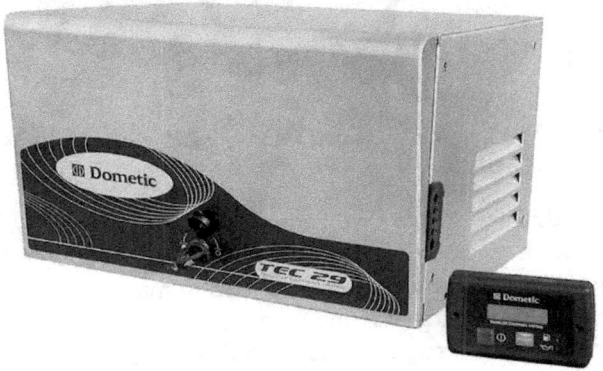

Figure 2.7. The Dometic TEC 29 petrol generator. Pic: dometic.com

A few generators (including the Honda inverter range), have a higher dc voltage and may even overcharge. Honda states (albeit confusingly) 'the output can charge batteries - but is not a battery charger'. The company recommends charging only to 50% - but that state of charge is hard for user to assess. With all such generators, it is far more effective and safer to use a high-quality mains-voltage charger plugged into the generator's 230 volt outlet. All such inverter generators work well with switch-mode chargers.

How noisy?

Most high quality inverter-generators' sound level is akin to that of people conversing in otherwise quiet surroundings. They can only just be heard running from with an RV, but may disturb campers in otherwise quiet surroundings at night. Technically, most have a noise level of 54 dB - 56 at 7 metres when running at about 75% output.

Diesel/LPG generators

Quiet diesel generators are rare and costly. Onan makes a good 5.5-kVA diesel unit suitable for large RVs and cabins. Onan and others also produce LPG-fuelled versions of their petrol-engined units.

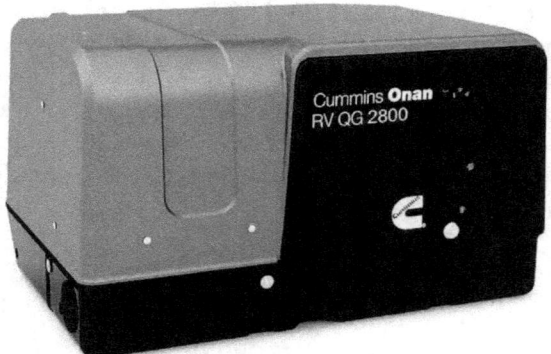

Figure 3.7. This Cummins-engined Onan QG 280 generator runs on LP gas. Specified noise level is 70 dBA at three metres. Pic: Onan.

The more costly generators have remote stop/start facilities. These can be interfaced to a solar or other regulator (or a BMS system) to operate automatically when the battery and load demands. It is, however, annoying to have a generator starting up automatically in the middle of the night. The complexity also complicates fault-finding.

Fuel-cells

Fuel-cells are very much the (electrical) energy source for future cabins and RVs.

A fuel-cell converts fossil fuels such as methanol, petrol, diesel, LP gas into hydrogen, thus enabling electricity to be produced electro-chemically. Unlike generators, there is no thermal burning process, no noise and next to no pollution. Fuel usage is roughly proportional to the load.

Figure 4.7. The EFOY Comfort fuel-cell outputs 6 amps at 12 volts. It is the size of a small suitcase and weighs 7.5 kg. Other sizes are available. Pic: EFOY.

The main small fuel-cell currently (late 2019) on the market is EFOY's range of methanol.fuelled units. They are of varying outputs (and also available to military specification). They originally cost $3500 upwards but now cost over twice that. Most are about the size of a jerry can, and weigh only a few kilograms.

EFOY fuel cells generate five or so amps upward (at 12 volts dc) and can be run 24 hours a day if required. A battery is required to cope with peak loads greater than the fuel-cell's typically limited low maximum available current. Because of its ability to store and deliver high current, a small lithium battery makes an ideal such partner.

These fuel cells are also an effective and silent back-up for solar. For most people, however, their initial price and high fuel cost currently rule them out as a major source of electrical power, however (and whilst this book has been suggesting this since 2006, and it has yet to happen) if/when their prices fall, fuel-cells are likely to wipe noisy generators off the face of this planet.

There is also an increasing possibility/probability that emissions legislation may eventually limit or prohibit using a road vehicle's alternator for any but its originally intended purpose. If that happens, fuel cells will be an acceptable alternative.

Chapter 8

Inverters

An inverter converts a typically 12-48 volt dc supply to 230 volts ac. Many cabin and RV owners buy one primarily to drive a microwave oven, then, having 1500 watts or more available, use it to run appliances like blenders and coffee grinders. If you don't need a microwave oven, or intend to run 230 volt power tools, you can run most small domestic appliances from a good 350-500 watt sine-wave inverter.

Inverter types

There are two main types of inverter output. These are square-wave, often and misleadingly called modified square wave (or 'simulated sine-wave'), or those that truly are sine wave. Sine-waves are that distributed as grid power, square waves are like multiple blocks of electrical energy.

Early square-wave inverters drove most appliances of the 1960-1970s, but were extremely inefficient. The later modified square-wave versions will drive most, but not all, electronic equipment. They may damage some (particularly laser printers) and can cause an annoying hum on radios, TVs and motors. They may cause electric motors to overheat.

Modified square wave inverters are still made. Quality ranges from acceptable to dreadful. There is little or no point in buying one as a high-quality true sine-wave inverter does not cost that much more and produces clean mains-equivalent power.

Today's best sine-wave inverters are also very efficient. The best waste less than 5% or so of the energy in the conversion. Some sense current draw and switch themselves to a very low current standby mode when they sense zero load. Others however remain fully working even if the load is tiny and draw a few watts constantly. This is hugely inefficient if the only load is (say) a 230 volt clock. A battery-powered clock is the obvious solution.

Transformer- based inverters

Apart from sine wave and modified square-wave, there are two main ways of designing and building inverters: transformer or switch-mode.

Transformer-based inverters are like silicon donkeys. They'll carry a heavy load briefly, a medium load for longer, and a comfortable load indefinitely. If you don't know about inverters (or donkeys), you are likely to choose one bigger and more costly than you really need. Transformer-based inverters are intended to work this way: even if truly overloaded they are still not damaged. As with donkeys, they stop working for a time to cool off. Even when running continuously at full load, a transformer-based inverter is unlikely to be overladen by the added short-term load of (say) a small blender or coffee grinder.

Most high quality transformer-based inverters maintain twice or more their rated output for some seconds, and 130%-150% for up to 30 minutes. These are not 'overloads' as such. The ability to work like this is simply a characteristic of an iron-cored transformer. The inverters' other bits are cheap so it costs very little more to provide that peak output.

This ability is needed for many power tools, water pumps etc that have electric motors that draw three or more times their running current for a second or so whilst starting. When assessing size, include only other appliances that are likely to be run at the same time, not those that just might be.

Adequate installation essential

Because a good transformer-based inverter can briefly supply two or more times its rated continuous load, a 12 volt, 1500 watt such inverter may draw 250 amps. This requires starter-motor size cable, and at least a 300 amp hour AGM, or lithium iron battery to feed it.

Whilst 12 volt 2000 watt inverters are made, most inverter makers recommend a maximum output of about 1200 watts for 12 volt inverters, 2400 watts for 24 volt inverters, and 4800 watts for 48 volt inverters. A 150 watt transformer-based inverter will drive a radio, TV, VCR, DVD, and small appliances, but not if several are in use at the same time. A 250-350 watt transformer-based inverter costs only a little more and will run small power tools for short periods, but 500 watts is more comfortable. A 230 volt microwave oven needs a 1500-1600 watt inverter, and more solar and battery capacity. Most washing machines run (on their cold cycle) from a 250-350 watt such inverter.

If more than 2000 watts is required (from a 12 volt supply), it can be done by using two inverters in parallel: only a very few can be used this way - see below (under Paralleling Inverters).

Switch-mode inverters.

Switch-mode inverters use a solid-state transformer-less technology. They are smaller, cheaper and very much lighter, but none has the 'overload' capacity of a good transformer-unit. Most, even those higher priced, are rated at only 80% of their claimed output if run for more than a few seconds. Care is required when buying as some are limited to 50% or even less.

Figure 1.8. Four paralleled Outback Power inverters.
Pic: Outback Power.

These inverters can be an excellent buy if overload capacity is not required but, if it is, they must be scaled accordingly. If overload capacity is needed, buy an inverter that is transformer-based.

Paralleling inverters

Some inverters (such as those from Outback Power - Figure 1.8) can be parallel connected to increase capacity. If you have this in mind make absolutely sure the units really do have this capability. Do not take a sales person's word for this. Only a very few brands of inverters can be used in this way.

Wired in - or freestanding

Many small inverters such as the Powertech unit (Figure 2.7) have one or more external socket outlets. Only appliances (including multiple power outlet boards) may be plugged into those sockets.

It is seriously illegal and extremely dangerous to have such inverters wired or plugged into fixed mains-voltage wiring as that bypasses essential safety devices. It may also result in mains voltage being impressed on a battery terminal resulting in a very real risk of severe electric shock.

Inverters intended for connecting into fixed mains-voltage wiring are made specifically for this purpose. Their secondary winding is fully insulated from the primary winding. In Australia all Jaycar such products have this protection.

Figure 2.8. Powertech 180 watt inverter has a 300 watt peak.
Pic: jaycar.com.au

Automatic load sensing - 'phantom loads'

As noted above, many small inverters are automatically actuated when a light or appliance is switched on, reverting to a low-energy mode when the last load is turned off. This useful behaviour can be foiled by so-called 'phantom loads'.

Many electrical lights and appliances have remote hand-held controls that enable them to be turned on and off. These units are convenient, but (unless the appliance is turned off at the wall outlet socket), continue to draw energy.

Those appliances made prior to pre-2014 and many from third-world countries draw a continuous 4-20 watts if switched off only at the appliance. If not switched off at the wall socket, appliances such as TVs may draw far more energy than in actual use.

This applies also to many 230 volt units that power (say) 12 volt LED lights. Their associated little black boxes (known also as 'wall warts') that plug into power outlets continue to draw some energy, unless switched off at the wall socket outlet. This is particularly an issue in cabin and RVs that have limited energy.

As noted above, such so-called 'phantom loads' may convince an inverter that an appliance is still on, preventing that inverter returning to stand-by. This results in a double loss: that of the energy drawn by each phantom load, plus the energy drawn by the inverter's internals through being kept working.

Some inverters can be set to respond only to loads larger than phantom loads, but this may prevent them responding to wanted small loads, like an electric razor unless (say) a light is turned on at the same time.

Manually switched inverters are still available, but it is only too easy to forget to switch them off after use. On the whole it is better to have an automatic load sensing inverter and, if adjustable (most are) spend some time setting it up.

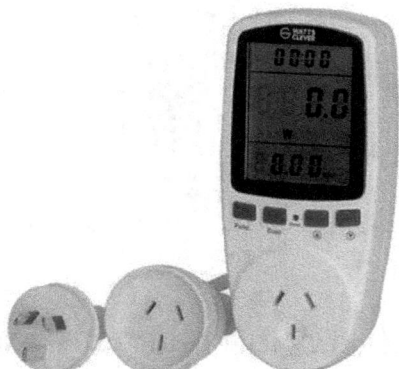

Figure 3.8. This energy/power meter from Jaycar plugs into the power socket via the cord. The unit to be tested is then plugged into the meter. Pic: Jaycar Electronics.

Post-2014 approved appliances are legally limited to such draw being less than 1 watt - but that's still an unacceptable waste in a cabin or small RV, let alone the typical 30 or so such loads in many homes.

To check actual energy draw, and whether or not an appliance draws energy if not switched off at the wall, use a 230 volt power energy meter (Figure 3.8). They are handy not just for RV use but also around the home. Good ones cost $20 upward.

Safety - a buying consideration

When buying an inverter, safety, as well as monetary cost, is involved. Some cheap inverters do not electrically isolate their 230 volt output from the 12 volt input. As noted above, this can result in a high voltage being imposed across one side of the battery.

High-quality inverters have their output fully isolated from their input. Reputable vendors explain why, and manufacturers make a point of clearly advising (in their sales literature and on the units) that their products are electrically isolated. As a general rule, unless there is specific evidence of electrical isolation, it is highly likely there is none.

Largely because of the above, there is a wide range of price for what may at first seem identical inverters, but are not. Good 150-500 watt electrically-isolated sine-wave inverters cost $1.00-$1.50 per watt. From thereon, price per watt progressively falls. Non-isolated inverters are a lot cheaper, but it is possible to replace dollars - not, however, yourself.

A really good-quality modified square-wave inverter will drive some loads, but far from all. Unless you know what you are doing buy only a good quality electrically isolated sine-wave inverter. There are only a few good inverter makers. Avoid hardware store specials, stay with known brands and buy only from reputable established vendors.

As emphasised on a previous page, most small inverters (those with in-built outlet sockets) must not be legally connected to such wiring. If they are, the resultant system is likely to be potentially (and often actually) dangerous. There is, in some circumstances, a real risk of electrocution.

Don't take a salesperson's word that an inverter is double insulated and/or electrically isolated. Few sales people have the technical background to understand what you mean. If an inverter is double insulated and/or electrically isolated it is all but certain that this will be made prominent in its promotional literature. If in doubt contact the manufacture.

Chapter 9

Refrigerators

The two main types of RV fridge are top-opening and door-opening. Those made specifically for RV use (and often used in cabins) have relatively efficient 12/24 volt compressors. These fridges originally drew less energy than most domestic fridges. There are however now several 230 volt ac domestic fridges of 220-250 litres that draw even less. One problem however is that many are 600 mm or more wide (and may not fit through an RV's typically 580 mm door). Whilst costly, Mielle has one that is 580 mm wide and is also ultra-efficient.

Three-way fridges, that run on LP gas, 12 volts or 230 volts when available, use the less-efficient absorption cycle but are still well worth considering as they hugely reduce the need for solar and battery capacity. They work very well but, as emphasised later in this Chapter they must be correctly installed.

Top or door opening?

Cold air sinks and flows out of a door opening fridge. If the fridge is opened frequently (as may happen with young children around) that loss can be substantially reduced by using high fronted plastic drawers. Keep the fridge full, but not tightly packed. Use plastic bottles full of water if necessary.

As cold air does not rise, none escapes when opening a chest fridge. Such fridges are thus more efficient particularly if frequently opened. Murphy's Law, however, results in the most needed items seemingly migrating to the least accessible place. This can necessitate a lot of cold items being taken out and repacked. Water condensing in the bottom needs frequent removal. The bottom also gets mucky.

This is less of a problem for chest freezers but can be for chest fridges and fridge/freezers as these are accessed more frequently.

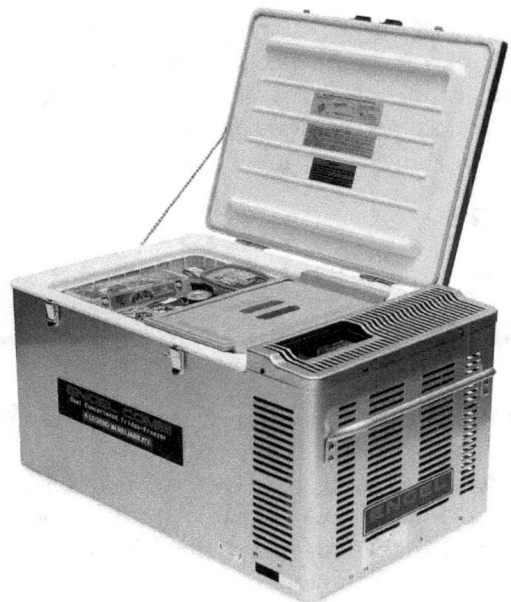

Figure 1.9. Engel 75 litre chest fridge/freezer. Pic: Engel.

Electric-only

Many electric-only fridges are made specially for caravans and motorhomes. A 75-80 litre chest fridge or fridge/freezer, however, really needs two 100 watt solar modules for that purpose alone.

Door opening RV fridges of 170-220 litres need three by 120 watt modules for reliable use in hot places. This is particularly so up north where it remains hot all day and night much of the time.

A few larger 12/24 volt door opening fridges are specialist-made, but most are converted standard domestic units.

Electric-only fridges are intended to run from the vehicle's alternator whilst driving, and from the RV's auxiliary battery and/or solar power when not. Most run well from solar.

With all fridges, energy consumption varies hugely with the nature of their installation, usage and ambient temperature. In practice very few are installed as their makers intended and, as a result, do not perform as well as they should. An electric fridge can almost always have its cooling ability enhanced and energy consumption reduced by correct installation. See Chapter 21.

Most electric-only fridge/freezers draw about 1 amp hour/day per litre of volume. Fridge-only units draw about 0.8 amp hours/day per litre. Large fridges use proportionally less. Fridges with variable speed compressor motors are claimed to draw up to 25% a day less.

Eutectic fridges

A eutectic reaction is one that uses a liquid, that when cooled, transforms into two solid phases at the same time at a specific temperature (rather than over a temperature range). It's a sort of 'now it's frozen - now it's not'.

Eutectic fridges uses a eutectic-like low-freezing-point liquid that freezes within tanks that form the walls of the fridge. That liquid is initially 'pumped-down' to that freezing point by running the fridge at its coldest setting for eight to ten hours. Once that is initially done, the eutectic reaction 'maintains that cold' for many hours. Because of this, such fridges typically need to be run for only one or two hours each morning and evening. In cold places a eutectic fridge may need only to be run for an hour or two every second day. In very hot places, 'pumping down' may need repeating from time to time.

Until recently (when used, in their pump-down mode) eutectic style fridges used less energy than a conventional chest fridge of similar volume. That energy saving, however, is not so much due to the eutectic principle per se. It is because a compressor motor's start up current is about twice that of its running current: energy is thus saved in such units because the compressor pump is not constantly cycling on and off in pump-down mode. The more recently introduced constantly-running variable speed (conventional) compressor fridges typically use 25% or so less energy than before. The eutectic units' main benefit now is that (in pump-down mode) they do not need to be run at night and are thus silent (a benefit for light sleepers in small cabin and RVs.

Intelliquip and Indel Webasto make such eutectic units.

Three-way fridges

*Figure 2.9. Dometic three-way fridge.
Pic: Dometic Australia.*

Three-way fridges run from the vehicle's 12 volt system whilst driving, 230 volts when available, and LP gas at all other times. They draw far too much current (10-25 amps) to run economically from an RV's solar energy.

Historically, three-way fridges had a poor reputation if used in warm climates. This is because most sold prior to 2000 or so were not designed to be used in ambient temperatures exceeding 25°C, but this was rarely known by sales staff, let alone buyers. Further, their performance was often degraded by seriously incompetent installation. If well installed they coped well enough most of the time, but could not handle semi-tropical, let alone tropical heat.

This situation changed in the late 1990s - when the European Union (EU) introduced a T-rating Standard for fridge cooling performance. This included four graded levels of ambient temperature performance.

T-rating

All EU-rated fridges have a compliance plate that contains the notation 'Climate Class'. To the right of that notation will be found any one of the letters SN, N, ST or T. These, respectively, are abbreviations of Sub Normal, Normal, Sub Tropical, and Tropical.

The SN and N are designed for ambients up to 32°C, ST up to 36°C, and T up to 43°C.

Do not confuse 'T-rated' with the term 'tropicalised'. All Dometic fridges sold in Australia have been marketed as 'tropicalised' following an upgrade to their specifications in the late 1990s, but the term 'tropicalised' is not the same thing as 'T-rated'. Dometic has never claimed it is, but far from all sales-people and owners seem aware of this. Unless the fridge has a EU compliance plate (Figure 3.9) that specifically states a fridge is 'T-rated' it is not.

The choice

Earlier editions of this book suggested that 110/120 litre fridges are the largest economically feasible for running mainly from solar. Since then both fridges and solar modules have become more efficient. The latter are also now very much cheaper.

Figure 3.9. This Miele (also sold as Liebherr) fridge-freezer is easily installed in most RVs as its 580 mm width passes readily through the typical RV door's 600 mm width. Most fridges are over 600 mm wide! Pic: Liebherr

Many medium-sized caravans and motorhomes settle for fridges of 170 litres, larger ones tend to have 220 litres or more. Camper trailers typically have 60 litres, with 80 litres as the realistic upper limit for solar located on the trailer or tow vehicle alone. Anything larger is likely to need solar modules on each vehicle.

Most users find electric-only fridges more convenient to use than three-way units. If solar module space allows, an electrical compressor-type fridge suits most applications.

Avoid having two or more fridges of small volume, rather than one of their combined volume. This is because a fridge's energy loss is mainly via its cladding. The area of that cladding is non-linear with the fridge's volume. As a result, two small fridges have far more than twice the area of cladding than one of the same volume. A few quick sums may surprise.

Three-way fridges cost more initially. Medium sized units use up to 0.4 kg of LP gas a day. Most owners find that an 8.5 kg LP gas bottle lasts about three weeks.

Eutectic fridges are efficient on their pump-down mode, but take all day to drop from ambient temperature to zero the first time they are used that way. This is not a problem for long term travellers, but can be inconvenient if the fridge is used only occasionally. They can be run on a low thermostat setting but they then draw much the same energy as other fridges of the same size and age.

The variable speed compressors now used in some conventional compressor fridges are quieter, but may still disturb light sleepers.

Correct installation

As stressed throughout this book, all fridges need to be competently installed, yet few are. If they are not, an electrical compressor fridge may still perform well enough, but cycle on more often and for longer - and thus draw a lot more power to maintain that performance.

Three-way fridges however cannot compensate that way. Unless properly installed, as shown in Chapter 21, they are unlikely to perform as they should. But, if installed correctly, post-2000 three-way EU-rated fridges really do work as specified.

Chapter 10
Lighting

Mains-voltage incandescent globes were banned from sale in Australia (in 2011), and many other countries. They produced more heat than light and, except for short-span use, were too inefficient to be driven from solar energy. They are still available in 12/24 volt form but as more efficient alternatives are now available there is little point in using them.

Regardless of the lighting source, a great deal of energy can be saved by painting walls white, or in a light reflective colour. Beige (particularly) and teak/walnut veneer suck light like a spectral vacuum cleaner, absorbing three or four times as much light as does white.

Switching

Internal lights are best switched individually from close to where they are located. External security lights need switching from two places. One switch should be readily accessible during the evening, the other from where it can be reached from the bed. Banging and coughing is usually due to nocturnal animals hunting down accidentally left-out Tim Tams or Camembert, but it is good to be able turn the outside light on to confirm that without having to get out of bed.

Halogen

Halogen globes will be banned from further sale as of September 2020. They are tiny incandescent bulbs filled with a gas that enables them to run at extremely high temperature (about 700°C). They are made in 12 and 24 volt form with energy draw from 10-50 watts. The 50 watt versions generate a great deal of heat and some require special large-pin holders. Care must be taken that the heat they generate can readily escape - this is a known fire hazard.

Figure 1.10. Fifty watt halogen globe.

These globes provide usefully focused light but draw far more power than the light emitting diodes (LEDs) described below.

Fluorescent lights

Fluorescent lights use about half the energy of halogens to produce the same amount of light, but most need 230 volts to drive them. The warm-white fluorescent tubes and globes have a pleasant glow similar to incandescent globes. Twelve-volt fluorescent lights are available but are 230 volt items with inbuilt inverters. As with halogen globes the era of fluorescent lights too is passing. As also explained below, it is now feasible and sensible to replace them by light emitting diodes that fit into the existing fluorescent fittings.

Figure 2.10. This is a 23 watt warm white compact fluorescent. Efficient and effective, but too bulky for small RVs.

Light emitting diodes

In 1907 Marconi laboratory's Henry Joseph Round noted that if about 10 volts is applied to carborundum (silicon carbide) crystal, it emits yellowish light. It was until the 1970s, however, that Fairchild Semiconductors managed to produce light emitting diodes at an affordable price and with a variety of types.

Light emitting diodes emit over four times more light per watt that incandescent lamps and are very much smaller. They cost more initially, but that is offset by their having a far longer life-span.

Originally, most LEDs provided a focused beam but many are now available that provide 3600. They are ultra-economic and especially effective for lighting discrete areas such as kitchens, cooking on campfires and reading. Providing the areas have light colours, reflected light provides effective background illumination. They are now readily available in a vast assortment of shapes, outputs and colour temperatures, and have all but replaced other forms of lighting in RVs.

The smaller LEDs are direct replacements for halogen globes. They are about 35 mm in diameter and fit into an MR 11 base with pins 4.0 mm apart. A larger version is 51 mm in diameter fits into an MR16 base and has pins 5.3 mm apart.

Both the MR 11 and MR 16 are fine for cabins, but less so for RVs as they rely on friction to keep then in place. They tend to fall out of basic fittings (particularly on rough roads), but fittings that retain them securely are readily available. There are also now many types of LED light fittings for RVs etc that have integral long-lasting LEDs.

There are also 230 volt LEDs that look much as the 12 volt versions but have GU-10 bases (larger two-diameter locking pins). Some are much larger, and with Edison screw or bayonet type fittings.

Whilst LEDs can be bought for typically a third of the retail price on many Internet sites, most of the ultra-cheap ones are very inefficient and have proved to have a short life span.

Colour temperature

Regardless of type, different light sources produce different coloured light. Candles burn with a yellow tinge, midday sun rays are slightly 'blueish'. These colours can be expressed as that which a certain material will produce when heated to varying temperatures - and referred to as 'colour temperature'.

Colour temperature is measured on the Kelvin scale, which is denoted by the letter 'K' or the word 'kelvin' after the number.

This primarily relates to LEDs and the various 'white' light they produce. Most are available in warm white (about 27000 K- 31000 K) and daylight white (50000-60000 K) and a few part-way between at 40000 K. Check this before buying as the 'daylight' type globes result can have an overly 'sterile' type of light that many find too harsh.

Light output

Until the advent of LEDs people somehow 'knew' the light output per watt as there was be next to no difference between brands of incandescent globes: for example a 60-100 watt globe was fine for most kitchens.

Figure 3.10. This 5 watt LED fits the standard (MR16) halogen globe holder - it has about the same light output as a 20 watt halogen globe.

LEDs however are very different. Light output (for the same wattage) varies: a high-quality 5 watt LED is likely to produce a lot more light than that does a 7 watt eBay cheapie - and last far longer.

Figure 4.10. Direct LED replacement for 230-volt incandescent Edison-screw base globes.

Furthermore, whilst most incandescent globes produced light over a virtually 360 degrees sphere, most LEDs have a confined beam of 15-1400. Because of that, and major differences in efficiency, to indicate their lighting ability, LED globes are now rated in units known as lumens.

Lumens

A lumen is a measure of the total amount of light emitted: in other words, the more the lumens the brighter the light.

The unit is also a measure a light source's energy efficiency. The more lumens per watt the higher the efficiency. (e.g. lumens per watt).

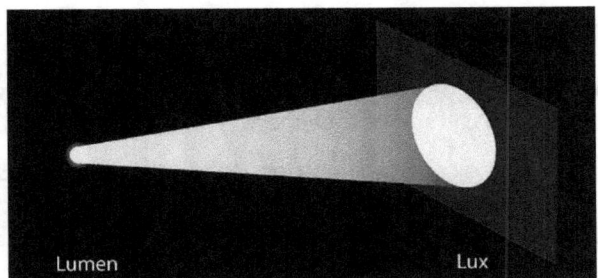

Figure 5.10. The difference between lumens and lux. Pic: original source unknown.

Lux

Lux is related to lumens but takes into account the area that is illuminated. One lux is equal to one lumen per square metre. The light level from a light source of (say) 1000 lumens over 1.0 square metre is 1000 lux.

Were that 1000 lumens spread over 10 square metres the average level would be 100 lux, i.e. lux takes into account the total area over which the light is spread (Figure 5.10).

Lux is that which most people will be primarily concerned about. Cabins and RVs need background levels of 50 to 60 lux. General tasks like reading and writing require about 350 lux. High precision work may need 500 to 600 lux.

Most LED makers disclose the lumens per watt. Light fitting suppliers are well aware of the common confusion regarding this and have live displays accordingly.

Light fittings for RVs

Any number and style of affordable LED light fittings for RVs are readily available from lighting stores and RV appliance suppliers such as Camec.

Many people still hand-craft LEDs lights using existing. An elegant example, made some years ago by Lawrie Beales, is shown below.

ReNew magazine (a quarterly available from most newsagents) has various project kits that enable large wattage LED light fittings to be built for a small fraction of the cost of commercial offerings.

Silicon Chip magazine carries associated articles and LED constructional projects. Components are available from Altronics, Jaycar, Oatley Electronics etc.

The choice

As noted elsewhere in this book, extra-low voltage wiring (e.g. 12-24 volts) is simple and easy, and you can legally do it yourself.

For lighting, my strong recommendation is for warm white LEDS. Their energy draw is so small that, if the cable is strong enough mechanically, it will be adequate to carry the required current. The readily available 1.5 mm² is electrical overkill but any smaller lacks adequate mechanical strength for RV use.

LEDs are thus a practical and convenient alternative in those only too many RVs where the existing cabling is thinner than desirable. You can always tell if such cabling is inadequate: lights dim discernibly when another light is turned on, or as an electric fridge cycles on and off.

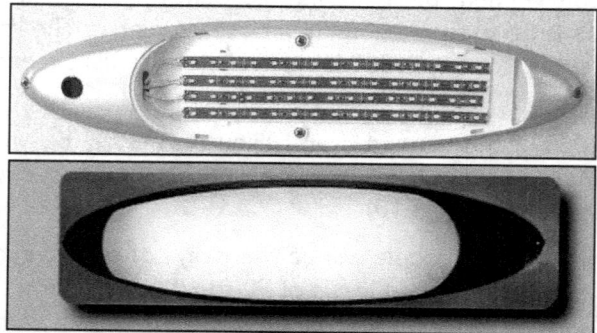

Figure 6.10. This elegant LED light fitting was crafted by Lawrie Beales. The internals (lower pic show the multiple LED lights in strip form that provide good and evenly spread illumination. Pix: Lawrie Beales.

Chapter 11

Water and pumping

Water pumping from solar is rarely a problem in RVs but can be for cabins if water is needed for irrigation. If pumping less than 1000 litres a day, use 12 volt or 24 volt pumps. These draw less energy than most 230 volt pumps, although high-efficiency 230 volt pumps (usually variable speed) are available.

It is essential to match pump type and size with the specific usage. Get this wrong with most mains-voltage centrifugal pumps and energy usage soars. Seek specialised advice from the makers or knowledgeable vendors, stressing that low energy draw is essential.

Water pumps for RVs

The most commonly-used pumps in camper trailers and small caravans are of small, in-line cylindrical form. These can be mounted above the tank's water level, but are likely to need a one-way valve in the pipe between the pump and the tank to ensure reliable operation. These pumps are reasonably reliable if used frequently, but tend to stick if not. It is advisable to carry a spare pump - or at least a spare pump diaphragm.

Make	Volts	Amps	Flow L/min	Pressure kPA (psi)
Flojet 4405	12/24	3.9 (12 V)	11	137 (20)
Flojet 4325	12/24	6.3 (12 V)	14	137 (20)
Jabsco 44010	12	4.0	9.5	133 (20)
Jabsco EF0612	12	6.0	12.5	133 (20)
Whale EF0612	12	3.9	7.0	212 (32)
Whale EF0012	12	4.2	10	212 (32)
Jabsco 36800	12	6.0	12.5	22 (20)

Figure 1.11. Typical performance including energy draw. Starting current is typically twice the running current, cabling needs sizing accordingly.

A more satisfactory (but noisier) pump is the diaphragm type. These can be mounted above the tank's water level, but may need a one-way valve in the pipe between the pump and the tank to ensure reliable operation. These pumps are reasonably reliable if used frequently, but tend to stick if not. It is advisable to carry a spare pump diaphragm.

Basic pumping systems require an associated switch to turn the pump on and off. This switch may be located close to the associated tap - or may be part of the tap itself. Such systems are cheap, simple and effective - but switches and water are poor companions - vital bits eventually corrode and stop working.

More friendly are systems that detect water pressure change. With these, pump water pressure is normally maintained in the piping. When a tap is opened, pressure falls. This fall is detected by a pressure sensing switch that starts the pump. The pump continues to run until the switch detects that the tap is closed and pressure is back up to normal. The switch then opens and the pump stops.

The pump continues to run until the switch detects that the tap is closed and pressure is back up to normal: the switch then opens and the pump stops.

Pipe resistance

Like electricity, but far more so, water resists moving through its conductor. A 12.5 mm water hose presents five times more resistance than does a 19 mm water hose, wasting energy accordingly.

With the commonly used irrigation hose, the 32 mm size has only one-third the resistance of 25 mm. The 40 mm size has only one-ninth. Whilst not an issue in an RV, this is a major issue for a cabin's irrigation system. Increasing pipe diameter may save far more than it costs by enabling the use of a smaller pump and less needed (ongoing) energy to drive it. Every right angle bend adds 5% so more resistance so, where possible, use wide-radius curves.

Pressure tanks

Basic pump pressure systems have the drawback that even tiny leaks, or changes in ambient temperature, cause the pump to restore lost pressure - often in the small hours of the morning. This can disturb sleepers as even vendor-styled 'quiet' pumps make quite a racket (the Shurflo WhisperKing is quieter than most).

Another issue is that high-tech washing machines and dishwashers turn the water on and off scores of times each washing cycle. Each on/off cycle causes the pump to start and stop. A pump's starting draw is at least twice its running current so energy use increases. Such pump cycling can be substantially reduced by including a pressure tank. This is a light steel or fibreglass tank that contains a strong balloon. This balloon is inflated to about 140 kPA (20 psi), i.e. just below a typical water pump's pressure switch's cut-in setting.

Pump cycling can be substantially reduced by including a pressure tank. This a (usually light steel or fibreglass tank that contains a strong balloon. This balloon is inflated to about 140 kPA (20 psi), i.e. just below a typical water pump's pressure switch's cut-in setting.

Figure 2.11. This pressure tank has the pump mounted sideways on its top. Pic: Water Worker.

When switched on, the pump progressively compresses the balloon until the pump's maximum working pressure is reached - about 315 kPa (45 psi). The pressure switch then opens and the pump turns off. The inflated balloon's pressure on the water then provides ongoing (totally silent) system pressure until the pressure switch detects it is necessary to switch the pump back on. This is typically when half of the tank's volume is used. A 20 litre tank thus provides about nine or ten litres of water between each pump cycle.

Given an adequate size tank, constant pump stopping and starting is considerably reduced. Water supply is silent between pumping cycles, water pressure is steadier, pumps last longer, people sleep more soundly. And the system it uses a fraction of the electrical energy.

Pressure systems are well worth considering for RVs spacious enough to accommodate their volume. They are also worthwhile for cabins, but not for irrigation systems where a constant flow is required.

Constant pressure pumps

Another constant pressure solution is the Shurflo Revolution pump. When water is drawn, the pump runs at full pressure but has an internal bypass within the pump from which water is drawn.

Figure 3.11. The Shurflo Extreme water pump. Pic: Shurflo.

If a long but only minor water draw is required, the pump nevertheless runs at full bore and thus consumes more energy than most pumps of this capacity. These pumps are compact and easy to install but thus best kept for uses where ample solar energy is available.

A second type of constant pressure pump (Figure 3.11) has a microprocessor controlled system that senses the required water flow rate and varies the pump's speed accordingly. The energy draw of this pump is more or less proportional to the volume of water drawn. Its maximum pressure is 420 kPA (60 psi), similar to a typical town supply.

Chapter 12
Computers and TV

Current (2020) higher efficiency 36 inch (90 cm) LED TVs draw 40-50 watts less than many 14 inch (36 cm) 12 volt TVs of older technology. The older ones in particular should be switched off at the power outlet after use as they continue to draw up to four watts only from the remote control. The same applies to masthead amplifiers, signal units used with satellite TV dishes and almost anything running from 230 volts. Turning the power off at the wall switch precludes that wastage.

Television

Well over half a TV set's energy is drawn by the screen and its drivers. That draw is proportional to the square of the screen's diagonal. LED screens usually draw the least.

Figure 1.12. Sony 36 cm LED TV. Pic: Sony.

The set-top boxes still used by some owners to enable analogue TVs to display digital programs draw excess energy as does the TV they drive. It is now far better to scrap both and buy a low-energy digital TV.

Computers

Many desk-top computers (particularly those made for gaming) draw too much energy for powering by typical RV solar.

The most practicable solution is a lap-top computer. Those with 17 inch (43 cm) screens also double as a small TV. Check their energy draw before buying as the more powerful ones draw 50-70 watts.

All computers need to be switched off at the wall outlet after they are shut down as they otherwise continue to draw a few watts.

Inkjet printers draw only a watt or so whilst idling. Older laser printers draw up to fifteen watts. Both use far more whilst printing (laser printers draw up to 1000 watts). This is not a problem for the odd letter, but becomes one if needing to print downloaded reference material.

Housing equipment

Neither computers nor TVs need sprung mounts, but are damaged by rattling around. Fix them rigidly, or pack them so they cannot move. The set-top boxes (used to enable pre-digital TVs to receive digital programs) are known to suffer damage unless securely located.

Be wary of wall-mounted TV brackets. Rough roads and corrugation cause the bracket to vibrate and possibly fracture, or the screws to pull out of the wall. Then down comes bracket, TV and all.

Both TVs and computers run just fine from sine-wave inverters, but some modified square-wave inverters cause problems. Few if any will run laser printers. Some are likely to damage them.

Chapter 13
Communications

NextG, 3G, 4G and now 5G services provide mobile phone and wireless broadband coverage for about 97% of Australia's population, but as of late 2020 sparsely populated areas of the Northern Territory and inland western Australia are not yet covered: nor is most of the inner land mass, or the dirt tracks that run through them. The network is still being extended and further developed, but as the remote inland population is so small, total coverage is unlikely.

For most RV users this coverage is just fine. It covers all major routes and many minor ones.

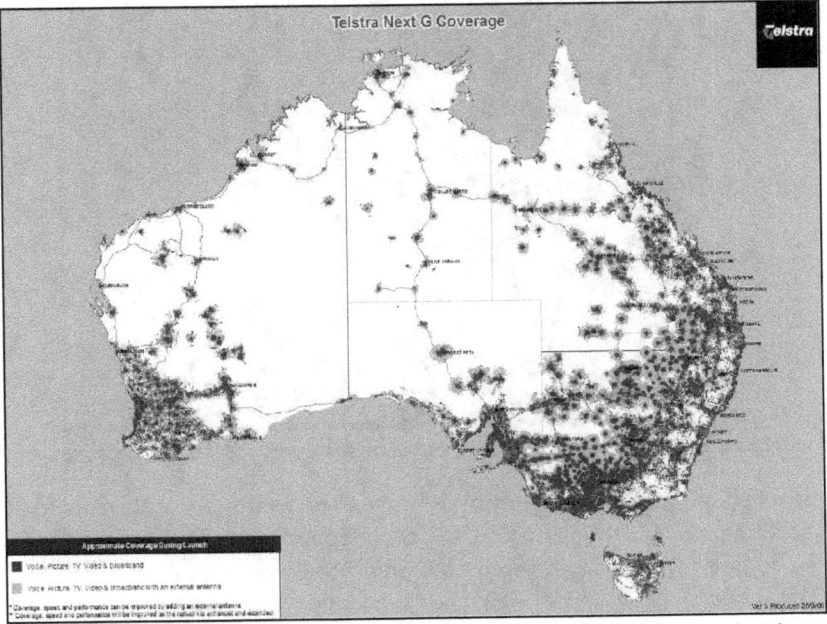

Figure 1.13. Testra's phone coverage (2019). Optus's is slighty greater, but less so in some outback areas. Pic: Telstra Australia.

In Australia's outback, there is usually mobile phone coverage within about 30 km of Aboriginal communities, but not otherwise.

The usable range can be extended by using an external antenna that needs connecting to the phone. There are various kinds and generally rated in dBi. On this scale the higher the rating the higher the signal strength the antenna will achieve: + 1 dBi is 12% higher, + 2 dBi is 58% higher, + 3 dBi is 100% higher and 6 dBi is 300% higher.

(If intending to travel extensively in Australia's more remote outback areas the only satisfactory means of communicating is via satellite phone - and covered later in this Chapter.)

Yagi Antennas

High gain (directional) Yagi antennas provide exceptional performance. They are not suitable for mobile use but ideal for remote cabins.

Charging the phone

Figure 2.13. Smartphone installed in a motorhome. Pic: Laurie Hoffman.

Mains-voltage phone chargers draw only a few watts but unless used at the same time as other appliances, will keep an inverter drawing continuous power. If feasible, avoid charging mobile phones overnight: older mobiles recharge in an hour or two, while smart phones take a bit longer, but the inverter and power supply continue to use energy for the rest of the night.

Figure 3.13. Charging a mobile phone for a USB adaptor that plugs into a standard vehicle 12 volt cigarette lighter outlet.

An efficient method, where voltage permits, is to charge the phone via an adaptor from the RV's 12 volt supply - Figure 3.13.

In-vehicle mobile phone kits have efficient inbuilt 12 volt chargers and also connect readily to an external antenna.

Satellite phone

This is *by far* the most reliable form of outback communication, but requires an unobstructed line of sight between the satellite phone and the satellite. Australia's is low down and to the northwest (it is over Papua New Guinea).

Whilst initially far larger and heavier (pic right) mobile satellite phones are not that much larger than normal mobile phones. Most hand-held mobile satellite phones have a short extending antenna, but some have provision for plugging in a high-gain antenna for use in areas of marginal reception.

Figure 4.13. Author's wife (Maarit) using her solar charged Iridium satellite phone - in Western Australia's mostly unpopulated Kimberley.
Pic: Author.

Satellite phones use more energy than do normal mobiles, but can be powered via plug-in converters from 230 volt inverter power, or alternatively from 12 volts dc as shown in Figure 3.13. They can be readily be run from solar.

High Frequency radio

HF radio has limited value for general communications. It provides coverage only between HF transceivers across limited and far from reliably predictable areas. There is also a 'skip distance' (typically 70-100 km) from the transceiver across which communication is only rarely possible.

These radios use 20-100 watts when transmitting but less when receiving. This is not serious for short calls, but becomes so if the set is left on 'stand-by'. An alternative is to switch the unit on only whilst driving, or at agreed times, allowing unscheduled incoming calls to be intercepted by private HF service message banks.

The best known such service is currently provided by HF-Tel (a division of the Australian National 4WD Radio Network) allowing easy to use 'short-term' or 'long-term' radio-telephone facilities exclusive for VKS-737 Radio Network subscribers.

Direct Dial Radio-Telephone calls to any fixed or mobile telephone number within Australia is available for a fixed price per month and includes 10 minutes worth of calls.

Whilst still in vogue, in RV Books opinion HF radio is now best left to enthusiasts for whom it is part communications but mostly hobby.

Personal locator beacon (PLB)

It is only to easy to become lost in Australia's less populated areas. This is particularly so in dense bush, that to all but bush dwellers seems uniformly identical.

Do consider carrying a Personal Locator Beacon (PLB). When activated, it transmits a signal that's picked up by satellites orbiting the Earth. This alert is then relayed to emergency services. The distress signal is strong and able to operate despite tree cover and heavy cloud etc, but if feasible move to an open area as high as possible with a clear view of the sky.

Figure 5.13. Personal Locator Beacon (PLB).
Pic: adventure safety.com.au

The ground rescue crew or helicopter crew go to the coordinates that the PLB transmitted (or uses radio direction finding to home in on the beacon).

Only use a PLB if you can't talk directly to emergency services and urgent assistance is required.

CB radio

CB radio is still very much in use, and is now extended from the previous 40 channels to 80 channels. Its range depends on the nature of the terrain, from line of sight to 50 km or more.

Short-range transceivers are also available. Caravanners in particular use them to advise their partners when reversing into tight places, and also to communicate with long-distance truck drivers. This can be done by having a pair of transceivers, or having just one used in conjunction with the vehicle's own CB radio.

In-car CB radios are normally muted to remain silent between signals. It is easy to forget they are still on when you leave the vehicle, resulting in a flat starter battery (but readily recharged if you have solar modules). Avoid this by wiring the CB radio via the ignition switch.

Useful Telstra links:

telstra.com.au/mobile-phones/coverage-networks/our-coverage/coverage-search/

Telstra mobile broadband coverage map: telstra.com.au/mobile/networks/coverage/broadband.html.

For Telstra Mobile Broadband devices (contract and pre-paid): telstra.com.au/bigpond-internet/mobile-broadband/

telstra.com.au/mobile-phones/coverage-networks/our-coverage/coverage-search/ telstra.com.au/mobile-phones/nextg-network/

Chapter 14

Scaling the system

Previous chapters of this book provide an overview of what is and what is not economically feasible with solar in cabins and RVs. This section shows how to work out what size and how many solar modules and batteries you need for virtually any sized system, from a single LED in the back of a VW Kombi, to systems for large cabins.

The starting point is to list what you believe you need, do some basic sums, and see how it all works out. Examples in Chapter 15 describe actual proven systems. If your proposed energy usage is a lot higher than a system described there that seems close to your needs, you may need to reconsider your plans as your system will otherwise cost more than average, or not be feasible through lack of space for solar modules.

Table 1.14 (below) lists typical appliances that are practicable to run from solar in RVs and cabins.

In Table 2.14 enter the watts used by your own appliances (particularly of an electric fridge). If unknown, use the typical consumption listed.

If usage is shown in amps for 12 volt devices, multiplying by 12 converts it to watts. For 24 volt systems multiply by 24.

* A microwave oven's wattage rating relates to its heating capacity, not to the power consumed in generating that heat. A typical '800 watt' microwave oven draws about 1350 watts, or about 1500 watts if via an inverter, allowing for both the inverter and charging losses.

Table 1 - Typical consumption - in watts

Appliance	Watts
Cassette/CD player	30
Coffee grinder	75
Computer (laptop)	20-30
Computer printer	70
Fans (12/24 volt)	10-25
Fridges - see text on pages	33-35
Lights 12-volt LED	3-5
Lights 12-volt halogen (each)	10-20
Lights - 240-volt fluoro (each)	8-18
Macerator	300-350
Microwave oven ('800 watt') *	1500
Mobile phone charger	5-10
Radio	15-20
Sewing machine	75-100
Stereo	50-60
TV (10-14 inch)	20-40
TV (16-20 inch)	50-80
DVD	30
Washing machine (on cold water)	200
Water pump (12/24-volt)	50

Table 1.14. Typical energy consumption in watts. Table: solarbooks.com.au.

Table 2.14

1. In column A, list all lights and appliances.

2. In B, enter the wattage, or the total wattage if more than one device (such as lights) will be used at the same time. For anything driven via an inverter made prior to 2012 either add 15% -or (and preferably) replace it by a high-quality post-2014 unit. This applies also to appliances: most made from 2014 onward are far more energy efficient.

3. In C, enter the hours each is used daily. Use likelihood, not rare maxima.

4. Multiply each entry in B by the respective entry in C.

5. Enter the amount totals in D.

6. Total all the entries in D and add 15% for charging/recharging losses.

The total shown in 'D' is the probably daily draw and is the minimum amount of energy that needs (on average) to be generated each day to counterbalance that used.

As emphasised in Chapter 3, solar modules typically produce about 70% of the industry-claimed wattage output. If cost and space permits, however (as it often does with cabins and large motorhomes) it is well worth assuming only 50% to allow for freak weather conditions and to speed battery recharging after heavy use. Doing this is not always possible but where it is that extra solar capacity will provide a truly robust system.

Column A	Column B	Column C	Column D
Device/s	Watts	Hours/day	Watt hours/day
Total of Column D			
As in item 2 (above) add whatever necessary for inverter losses			
Total - the most probable watt hours/day energy draw			

Table 2.14. Use this to list all details of lights and appliances.

Two solar approaches

There are two main, but different, approaches to using solar in RVs and cabins.

The first is to use solar to generate as much energy as you use, plus a margin to allow for losses, and for short periods of unusually low solar input. Where space allows, a margin of 30%-50% more solar capacity is well worth the now relatively low extra cost. When calculating this, assuming that solar modules produce about 50% of that claimed (they actually produce about 71%) allows for voltage-drop along wiring and inevitable battery charging/discharging losses. This is the minimum solar capacity that will provide for most typical cabin and RV usage. Ideally, install all you can!

The second approach is to supplement (less) solar input by alternator-charging batteries whilst driving, or from a battery charger powered by a generator's 230 volt output, and grid power when available. The following text and the solar output (Table 3.14) are pertinent to both. This approach will be needed if planning to travel in southern Australia (especially Tasmania) in winter as solar input is usually too low for its use alone.

How many modules?

The method and data for calculating the solar capacity recommended in this book assumes (as does the solar industry) that typical ambient temperature will be mainly 20°C-25°C. If it is higher, you will need about 5% more solar capacity per 1°C increase.

The Peak Sun Hour Maps (Figure 1.14) show the most likely maxima and minima for Australia in mid-January and mid-July. In most places, the solar irradiation varies linearly for the months in-between. From these maps you can ascertain the average daily Peak Sun Hours (PSH) most likely experienced.

Most people who use an RV for holidays, assume 3.5 to 4.0 PSH. For those who intend to travel year around, 3 PSH should suffice for all but the lower parts of Australia's south during mid-winter. There, a generator is usually needed as that area typically has only 1 to 1.5 PSH in June and July.

If you assume higher Peak Sun Hours than above, and particularly if you have a fridge-freezer, consider carrying a back-up generator or a fuel-cell (Chapter 7). Or use a three-way fridge running on LP gas (whilst camping) to slash electrical usage (Chapter 9).

From the previous details you already know how many watts you are likely to use each day. Multiplying that by the relevant number of Peak Sun Hours per day gives the bare minimum watt/ hours that you need on average to generate each day. As emphasised above, in practice you really need more.

Table 3.14 shows the most probable average daily output of horizontally-mounted solar modules.

Table 4.14, and primarily applicable to cabins, shows the most probable average daily output of modules mounted at the optimum fixed tilt angle. Figure 1.19 (Chapter 19) shows how to estimate that angle. This angle is not critical, variations of up to 10% either way make little difference to daily input: there are only minor losses in having horizontal solar module mounting (and some gain in many parts of Australia except during winter). From thereon it is simply a matter of selecting whatever combination of solar modules is required.

Mounting is eased by using identical sized modules, but it is technically fine to connect say, a 40 watt and a 100 watt module of similar voltage in parallel (plus to plus, minus to minus) to increase amp hour capacity - in this example, to 140 watts.

Supplementing the battery

With the exception of the big motorhome in the example systems (Chapter 15), all the systems described there are substantially or wholly electrically self-sufficient. It is however readily possible to design the system such that solar energy only supplements that used. In other words, the battery still runs down, only later, and will then need generator or grid power recharging.

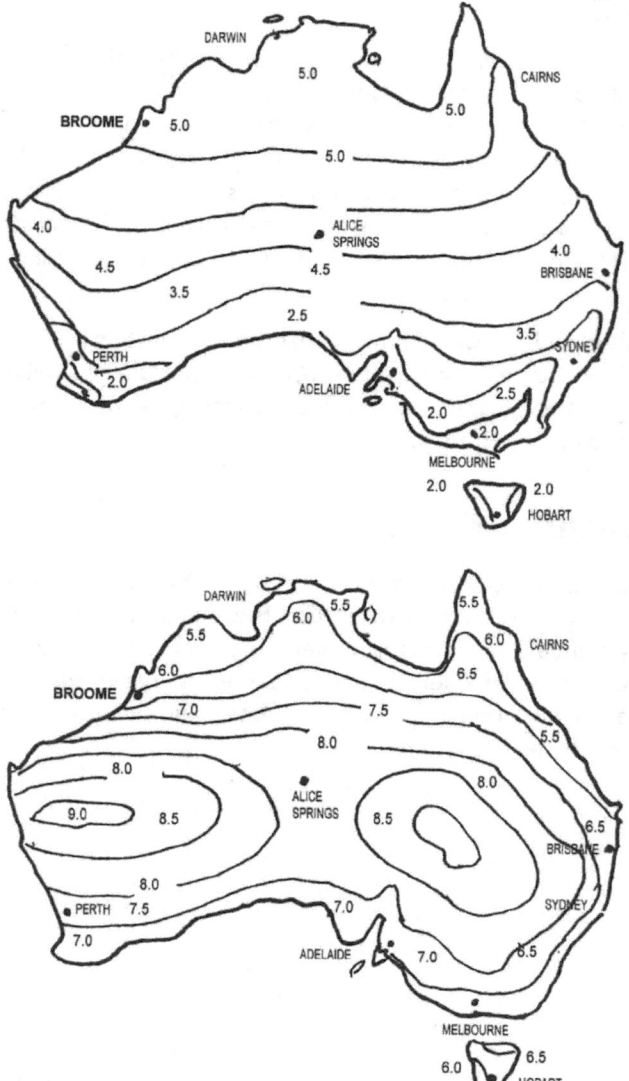

Figure 1.14. The upper Peak Sun Hour Map shows typical average irradiation for mid-July, the lower one for mid-January. Irradiation is more or less linear in between. To assess the probable daily input, take the data shown for your area of interest and work on 70% of the nominal claimed output of most solar modules. Pics: solarbooks.com.au

Solar output - modules flat

Modules: 64 watt	1	2	3	4	5	6	8	10
3 PSH	120	240	360	480	600	720	960	1200
4 PSH	160	320	480	640	800	960	1280	1600
5 PSH	200	400	600	800	960	120	1600	2000
6 PSH	240	400	720	960	1200	1440	1920	2400

Nominal 64 watt module

Modules: 80 watt	1	2	3	4	5	6	8	10
3 PSH	150	300	450	600	750	900	1200	1500
4 PSH	200	400	600	800	1000	1200	1600	2000
5 PSH	250	500	750	1000	1250	1500	2000	2500
6 PSH	300	600	900	1200	1500	1800	2080	3000

Nominal 80 watt module

Modules: 100 watt	1	2	3	4	5	6	8	10
3 PSH	187.5	375	560	750	935	1125	1500	1875
4 PSH	250	500	750	1000	1250	1500	2000	2500
5 PSH	310	620	935	1240	1550	1860	2480	3100
6 PSH	375	750	1125	1500	1875	2250	3000	3750

Nominal 100 watt module

Modules: 120 watt	1	2	3	4	5	6	8	10
3 PSH	225	450	675	900	1125	1350	1800	2250
4 PSH	300	600	900	1200	1500	1800	2400	3000
5 PSH	375	750	1125	1500	1875	2250	3000	3750
6 PSH	450	900	1350	1800	2250	2700	3600	4500

Nominal 120 watt module

Table 3.14. *Most probable output in watts of typical horizontally mounted solar modules.*

Solar output - modules angled

Modules: 64 watt	1	2	3	4	5	6	8	10
3 PSH	135	270	405	540	675	810	1080	1350
4 PSH	180	360	540	720	900	1080	1440	1800
5 PSH	225	400	675	900	1225	1350	1800	2250
6 PSH	270	540	810	1080	1350	1620	2160	2700

Nominal 64 watt module

Modules: 80 watt	1	2	3	4	5	6	8	10
3 PSH	170	340	510	680	850	1020	1360	1700
4 PSH	225	450	675	900	1125	1350	1800	2250
5 PSH	280	560	840	1120	1400	1680	2240	2800
6 PSH	340	680	1020	1360	1700	2040	2720	3400

Nominal 80 watt module

Modules: 100 watt	1	2	3	4	5	6	8	10
3 PSH	210	420	630	840	1050	1260	1680	2100
4 PSH	280	560	840	1120	1400	1680	2240	2800
5 PSH	350	700	1050	1400	1750	2100	2800	3500
6 PSH	425	850	1275	1700	2125	2550	3400	4250

Nominal 100 watt module

Modules: 120 watt	1	2	3	4	5	6	8	10
3 PSH	255	510	765	1020	1275	1530	2040	2550
4 PSH	340	680	1020	1360	1700	2040	2720	3400
5 PSH	425	850	1275	1700	2125	2550	3400	4250
6 PSH	510	1020	1530	2040	2550	3060	4080	5100

Nominal 120 watt module

Table 4.14. Most probable output in watts of solar modules facing north and tilted at latitude angle.

Assume you have a 200 amp hour battery that is initially charged to 90%, i.e. 180 amp hours - or (as amp hours times volts equals watt hours) 2160 watt hours. Discharging to 50% remaining, as battery makers recommend, leaves an available 90 amp hours - that is about 1080 watt hours.

If you use 500 watt hours/day, you can thus stay on site for two days on battery power alone. If you wished to stay on site for six days (requiring 3000 watt hours) - here's what you could do. There is an initial 1080 watt hours available from the battery: that is 180 watt hours for each of six days - 320 watt hours short of that which you need. Assuming 3 PSH/day and 70% efficiency that deficit can be supplied by two 80 watt solar modules (100 watt modules are often the same price.) Increasing solar capacity to two 120 or 130 watt solar modules will enable you to stay on site indefinitely, and with a well-charged battery for back-up during overcast conditions.

In essence if you add enough solar to stay on site for six or so days, it needs only a little more to be able to stay there (electrically) as long as there is sufficient sun. It is also now practicable because solar capacity has become so cheap.

Supplementing solar energy

Where you have even occasional short-term high electrical loads, total self-sufficiency becomes less practical. Here, it makes more sense to scale the system as above and use a quiet generator for those loads. Note however that most national parks and some caravan parks ban all generators. In these, consider using a fuel cell connected across a 50 Ah battery to cope with energy draw higher than a fuel cell's usually limited 5 amp or so maximum output.

Winter use

As noted previously, if you intend to spend time in southern areas such as Tasmania or around Melbourne during June-August, where there is usually less than 1.5 PSH, it makes little economic sense to scale the system for so little solar input. Recharge the battery bank via a small quiet generator driving a 230 volt charger and or via a fuel-cell (if/when costs afford). Or avoid such areas during mid-winter.

Electrical self-sufficiency

Electrical self-sufficiency can be achieved silently and pollution free and, if competently designed and installed, is ultra-reliable. It costs more initially, but batteries last much longer. Less frequent replacement goes part-way toward the one-off cost of more solar capacity.

Many RV and cabin solar systems are close to self-sufficiency (or achieve that) much of the time, but if, on average, there is any deficiency at all, that system must eventually run out of power.

Figure 2.14. Self-sufficiency as defined by Mr. McCawber (Charles Dickens): "Annual income twenty pounds, annual expenditure nineteen nineteen six, happiness happens. Annual income twenty pounds, annual expenditure twenty pounds ought and six, result misery." Solar is like this too.
Pic: original source unknown.

Never cut back on solar capacity.

Never have more battery capacity than you can fully charge by midday on almost all days of the year. If the energy is not captured it is not there to store.

The cartoon (Figure 2.14) says it all.

Chapter 15

Example systems

Two of the seven following examples describe systems originally built by readers of earlier versions of this book - when solar capacity was over three times today's price - but (as described) later updated. Those are included because many such systems are still in use and those examples show what is feasible and at what cost. Three are of systems designed more recently. One or another is likely to be similar to what you have in mind. If it is, check your own calculations against the example as a guide to ensure that you are on the right track.

Example 6 is included to show what can go seriously wrong in the event of a major misunderstanding between client and supplier, or ignoring expert advice.

If you find that your estimated energy usage is much higher than the closest example shown it may pay to revise your system as it is otherwise likely to cost more than average, or not perform as hoped.

Watts the matter

As elsewhere in this book there is an unavoidable complication - in that battery makers quote electrical energy in terms of volts and amps, but most solar and appliance makers use watts.

To convert volts and amps to watts, multiply amps by the system voltage (e.g. 12 or 24).

To convert watts to volts and amps, divide watts by the system voltage.

Please refer also to Chapter 3 re wattage claimed for solar modules to avoid incorrect assumptions

- unless the solar industry current practice (of overstating the most typical solar module output) is understood and taken into account. Alternatively assume that the most likely output will be about 70% of that seemingly claimed. Either way it is essential to work from the output the modules actually produce, rather than what can seem to be claimed.

The examples systems here all use 12 volt solar modules of up to 120 watts each. Some use combinations of different wattage modules that best fitted the space available. There is nothing rigid about this.

To increase capacity you can parallel modules of the same voltage but of different wattage output.

To increase voltage (e.g. to 24 or 48 volts) modules can be connected in series but all must be of the same current output and of the same type. This is often done with stand-alone home and property systems as its enables thinner power cables to be used without increasing voltage losses. Unless otherwise stated, the example systems described were intended by their owners to be capable of running indefinitely whilst away from mains power.

Example 1. Small cabin

Column A Device	Column B Watts	Column C Hours/day	Column D Watt hours/day
Fluro Light (2)	12	3	72
Radio	15	1	15
Portable TV (10 inch)	30	2	60
Water pump (12-volt)	48	0.15	8
Total of Column D			155
Plus 15% for system losses			23
TOTAL			**178**

This is a basic but adequate system originally built in 2002 for two people who spend most weekends from October through March in a small cabin a few kilometres north of Byron Bay (NSW).

The cabin has a three-way fridge that only runs on gas, a small 12 volt TV, a battery-operated radio that charges its NiCad cells from the 12 volt solar system, a couple of 12 watt compact fluorescent lights, and a Shurflo water pump. The average draw was 178 watt hours/day (approximately 15 amp hours/day).

At a realistic 4.0 Peak Sun Hours, the required total of 178 watt hours/day was comfortably supplied by a tilt-angle mounted 80 watt solar module producing about 225 watt hours/day. Solar output was handled by a 10 amp regulator.

An 85 amp hour gel cell deep-cycle battery provided three/four days usage with little solar input and up to seven days if TV watching was curtailed. The gel cell battery enabled the system to be left unattended during the winter months.

As long as that battery was left fully charged and isolated from the system, it still retained most of that charge at the end of winter. That original battery lasted for seven years and was replaced by a 100 amp hour AGM in 2010.

Solar capacity was increased by 30% (in 2015) to assist during periods of little sun, and also to extend battery life. The latter worked well as that 2010 battery is still (2020) more or less working, but now has less than half its original capacity and needs replacing. (The battery industry by and large sees a battery as at the end of its life when its capacity is 80% of the original).

This example would suit virtually any usage where low price and simplicity is paramount. It is also an ideal system for a basic camper trailer.

As with all the example systems described, savings can be made by using LED lighting, or having more and brighter lighting for the original consumption.

There is a lot to be said for basic systems like this. Ultra-simple, very reliable and with an electrical energy draw that is so low that battery capacity can cope with long periods of little sun.

Workings for above	
Total daily energy demand (+10%) approx. 171 Wh/day =	15 Ah/day
Average solar input at 4 PSH (1 X 80 watt angled module) is approx. 225 Wh/day =	18.75 Ah/day
Battery (AGM)	100 Ah

Example 2. Campervan (40 litre electric fridge)

Column A	Column B	Column C	Column D
Device	Watts	Hours/day	Watt hours/day
Fluro Light (1)	12	3	36
Radio	15	1	15
Portable TV (10 inch)	30	2	60
Water pump (12 V)	48	0.25	12
Chest fridge (40-litre)	36	8	288
Total of Column D			411
Plus 15% for system losses			62
TOTAL			**473**

Workings for above (as at present)	
Total daily energy demand is 473 Wh/day =	39.5 Ah/day
Average solar input (3 X 100 watt modules = 630 Wh/day)	52.5 Ah/day
Batteries (two by 75 Ah)	150 Ah

This example was built by a young couple (in 2010) for an extended low-budget trip around Australia in a converted 1974 VW Kombi. It is similar to Example 1 excepting that the three-way fridge is a 40 litre, 12 volt chest-opening Engel. Even though relatively small and electrically efficient at that time, this fridge almost tripled current draw, necessitating correspondingly larger solar and battery capacity. There was also a second 12 watt fluorescent light that was used only occasionally.

The owners sought to use solar-only virtually everywhere at all times. The system was scaled accordingly. Tasmania and the Eyre Peninsula were visited, but outside the three mid-winter months.

At a realistically conservative 3.0 Peak Sun Hours, the owners chose to use three flat-mounted 100 watt modules. These generated a comfortable 630 watt hours/day. A 20 amp Plasmatronics solar regulator that has inbuilt monitoring facilities was included.

The original 150 amp hour deep-cycle lead-acid battery capacity provided two to three days use with zero solar input. In 2017 the owners reported that the system had never been anywhere near running out of power and that the original batteries appeared to still be in excellent condition. A later check, however, showed their actual capacity had dropped to only 120 amp hour so they were replaced by AGMs of the same 150 amp hour capacity.

Again, a preferred type of system: economic, simple and reliable. It provides what most people need for basic camping and cabin living comfort but it would be worth increasing solar capacity to about 450 watts as that will provide more energy on overcast days.

Example 3: Small motorhome (120 litre electric fridge)

Column A Device	Column B Watts	Column C Hours/day	Column D Watt/hours/day
LEDs (3)	7	3	63
Radio	5	1	5
Appliances (via inverter)	350	0.5	175
LED TV (16 inch)	30	3	90
Chest fridge (120 litre)	60	10	600
Water pump (12V)	60	0.25	15
Total of Column D			948
Plus 15% for system losses			142
TOTAL			1090

This system was built by a couple who had previously travelled extensively in a small campervan but had now bought a 7 metre Toyota Coaster. They intended to winter only in warm sunny places and assumed a (rather optimistic) 4.5 Peak Sun Hours. As in example 2, they used a chest fridge but of 120 litres - using an extra 100 mm of insulation to limit daily draw to only 600 watt hours.

The system retained the previous 350 watt inverter to run a blender, a small computer via a 230 volt adaptor, and a new (in 2015) 16 inch LED TV. There are five LEDs lights (all 7 watts). Not all are used at the same time (the draw has been calculated accordingly).

The required 1090 watt hours/day energy is supplied by four flat-mounted 100 watt modules producing (a rather marginal) 1260 watt hours/day. There is a 30 amp solar regulator with monitoring facilities.

A 250 amp hour AGM battery bank (often discharged by 50%) provides a little over three days comfortable usage with next to no solar input.

This system has worked well so far but really needs 50% more solar (for which there is ample roof space) to ease the load on the battery when solar input is low.

Workings for above	
Total daily energy demand is 1090 Wh/day =	91 Ah/day
Average solar input 4.5 PSH (4 X 100 watt flat-mounted modules =	1260 Wh/day) = 105 Ah/day
Batteries (two by 125 Ah AGM)	250 Ah

Example 4. Average caravan/motorhome

Column A	Column B	Column C	Column D
Device	Watts	Hours/day	Watt/hours/day
LEDs (5)	7	3	105
Radio	15	1	15
LED TV (16 inch)	30	3	90
Fridge (170 litre)	80	12	960
Water pump (12 V)	48	0.5	24
Total of Column D			1194
Plus 15% for system losses			185
TOTAL			1380

The most typical RV usage is occasional weekend and long weekend trips, plus an extended tour every year or so. Given minor variations, this system and its usage is typical of tens of thousands of such medium-sized caravans and motorhomes. Such a system also suits many small/medium sized cabins.

As they never travel during the three mid-winter months, the owners of this medium-sized motorhome also originally assumed (an optimistic) daily 4.5 Peak Sun Hours. The required energy of 1380 watt hours/day was originally obtained from four by 130 watt modules (producing about 1640 watt hours/day at that assumed PSH). The MPPT solar regulator handled up to 40 amps. Battery storage is three 120 amp hour AGMs.

Although the owners originally decided against it, carrying a back-up generator is advisable for any solar-driven systems with a largish electric-only fridge. In this case it was truly pushing their luck as the solar capacity was not only marginal, but assuming 4.5 PSH is optimistic.

The owners found out (via a fridge full of warm beer and rotting food) that the solar capacity was indeed inadequate. My advice was that increasing solar capacity by 50% - 60% would result in less chance of dead batteries and a warm fridge, but that it would still be advisable to carry a 1000 watt inverter-generator plus a battery charger run from its 230 volt output; or stay as is and use a three-way fridge running on LP gas whilst camping.

The owners added 50% more solar and also bought a 1 kW generator plus 230 volt battery charger and have had no further problems.

Where, as in this example, there is intermittent use, battery choice and care is a major consideration. The AGM batteries used here hold well over half their charge for a year or more if initially fully charged. Unlike conventional lead acid batteries, they suffer no damage if less than fully charged.

A suggested change would be eventually to replace the batteries by LiFePO4s and use a 3 kW generator and the highest capacity battery charger they can find - as that will charge the batteries very much faster. Further, if the RV is stored (with the batteries 50% charged) those batteries will not require recharging until their next trip - and will suffer no harm.

Workings for above	
Total daily energy demand is 1380 Wh/day =	115 Ah/day
Average solar input 4.5 PSH (4 X 130 watt modules 1640 Wh/day) =	136 Ah/day
Batteries (3 X 120 Ah AGM)	360 Ah

Example 5: Large caravan

Caravanners have used three-way fridges running on gas for well over 60 years, but most pre-2000 or so lacked performance in any but mild climates and many were (and still are) badly installed. The 'T-rated' 'three-way' units do however work well (subject to competent installation) as is stressed throughout this book.

These fridges can work almost independently of electrical input. If solar is used it is readily feasible to cut consumption by not using any major electrical energy gobbler, such as a microwave oven, thus making it feasible to design such systems without providing a large energy margin for extended cloud cover.

This unit is a six metre twin-axle caravan used for semi-permanent living and travelling, often staying on site away from mains power for several weeks at a time. The T-rated fridge is a 2012 unit the owners say works really well even on days that are over 40°C.

The owners spend most of their time north of the Brisbane/Geraldton line and opted for a (there) realistic 4.0 Peak Sun Hours. The required 705 watt hours/day was obtained from two flat-mounted 150 watt modules (producing up to 840 watt hours/day). A 20 amp regulator was adequate. The high draw of the microwave oven necessitated AGM batteries. The owners chose two by 100 amp hour but three would better cope with the high current drawn by that oven.

The typical energy draw is usually only 550 watt hours as it is rare to use all five lights at the same time, and the microwave oven is used only occasionally.

This example highlights the huge electrical savings if using a three-way fridge. The owners say that the fridge, on average, uses 0.35 kg of LP gas a day. An 9.0 kg bottle (holding 8.5 kg of gas) typically lasts about three weeks. LP gas is obtainable almost anywhere - but is very costly in outback areas (so the they carry a spare bottle. A move to LiFePO4 batteries could be worth considering when the AGMs are retired.

Column A	Column B	Column C	Column D
Device	Watts	Hours/day	Watt/hours/day
LEDs (5)	7	3	105
Radio	15	1	15
Portable TV (20 inch LCD)	40	2	80
Fridge (three-way)	N/A	N/A	N/A
Microwave oven	1300	0.1	130
Other appliances (via inverter)	350	0.5	175
Water pump (12 V)	60	0.5	30
Fan	20	3	60
Laptop computer	36	0.5	18
Total of Column D			613
Plus 15% for system losses			92
TOTAL			**705**

Workings for above	
Total (max) daily energy demand (705 Wh/day) =	59 Ah/day
Average solar input 4 PSH (2 X 150 watt flat-mounted modules 840 Wh/day) =	70 Ah/day
Battery (two by 100 Ah AGM)	160 Ah

Column A	Column B	Column C	Column D
Device	Watts	Hours/day	Watt/hours/day
Incandescent lights (6)	36	3.0	688
TV (42 inch)	170	3.0	510
Fridge (electric 500 litre)	300	14	4200
Microwave	1300	0.5	650
Washing machine	250	0.5	125
Water pump (12 V)	60	0.5	30
Fan	50	5.0	100
Total of Column D			6303
Plus 15% for system losses			945
TOTAL			**7250**

Example 6: Coach conversion (old 500 litre fridge)

This system shows just how badly things can go wrong when an ambitious project is attempted by someone with no prior knowledge of solar or RV electrics. It was worsened by insisting on retaining a (then) 25 year old energy gobbling fridge and incandescent lighting in an otherwise well self-converted coach.

The owner/builder reasonably (but wrongly) not only assumed that solar modules produce that seemingly claimed, but that they produce their claimed full output whenever/wherever the sun is shining. Totally misunderstanding the Peak Sun Hour concept, he assumed an input of 12 PSH/day year around. He thus installed six 100 watt solar modules believing they would produce about 7200 watt hours/day. The reality is about 1600 watt hours/day (133 amp hours/day) on rare good days, but more typically 1200 or so watt hours/day (100 amp hours/day).

Worse, not initially realising he needed a solar regulator, the solar output was connected directly across the then totally unloaded 800 amp hour bank of (fortunately second-hand) 12 volt lead-acid deep-cycle batteries. As no load was being drawn, the batteries progressively overcharged, boiling themselves dry a few weeks later.

The owner then realised he was way out of his depth. He bought an earlier version of this book and contacted me for advice, giving permission to publish his experience in exchange.

The problems were readily fixed. I advised taking the fridge (and his now useless batteries) to the tip and to install a new and efficient 220 litre compressor unit, fit LEDs into the existing light fittings and (as weight and space was not an issue) to have an experienced electrician install four 100 amp hour AGM batteries (totalling about 130 kg) plus a solar regulator.

Allowing for losses, the final energy draw was just over 2500 watt hours/day (about 210 amp hours/day). Or about 1850 watt hours/day (155 amp hours/day) if the microwave oven was not used.

The owner opted to retain the existing solar capacity and added a 230-volt, 3 kW Onan generator and 60 amp battery charger to keep the AGM batteries always above 50% charge. This conversion had a bad start but ended up working reliably.

Workings for the original	
Total daily energy demand 7250 Wh/day =	605 Ah/day
Average solar input 6.0 PSH (6 X 100 watt modules) 1600 Wh/day =	133 Ah/day
Batteries (now four 100 Ah AGMs)	800 Ah

Example 7. Big fifth-wheel caravan - (low energy)

This is an 11 metre fifth-wheel caravan used for permanent living. It has central-heating and a big TV, yet uses less electrical energy than many a camper trailer! A Dometic three-way fridge in the trailer runs permanently on LP gas.

Whilst the 793 watt hours/day (including 15% excess) could be comfortably derived from four 120 watt modules, the owners opted for six. Three are on the tow vehicle's cab, the other three on the roof of the trailer: (a cab-located 12 volt fridge/freezer, runs from the alternator and has solar to charge a 100 Ah AGM auxiliary battery used only for shopping. Its draw is not thus included). The combined 1350 watt hours/day is far higher than needed but enables each vehicle to be electrically independent. The battery bank can be similarly split.

Both systems are normally paralleled enabling the combined 480 amp hour battery bank to provides about a week's use with virtually zero solar input. Paralleled, the system runs happily at 2.0 PSH/day. The inter-vehicle feed is via a 35 mm² two-metre cable and Anderson plugs and sockets. A 30 metre 16 mm² cable interconnects the units when further apart. There are two separate solar regulators.

The towing vehicle is normally in the sun and the cable-connected fifth-wheel caravan in semi-shade. This provides adequate charging almost all the time. When used apart, the two sets of batteries charge unequally, but self-equalise over time once interconnected. The heater pump in the equipment list (above) is part of the rig's (diesel) central heating system. Whilst a generator is carried, it is used only for running big power tools.

This is an excellent and versatile (albeit costly) system. Some 160-180 watt/hours could be saved by replacing the fluro lights by LEDs; and by replacing the TV by an LED unit (of up to 36 inch). If spending a lot of time in the tropics, it might be better to locate four modules on the towing vehicle and two on the trailer, enabling the trailer to spend more time in the shade. We used a generally similar but more modest concept for our Nissan Patrol and TVan - Chapter 29.

Column A	Column B	Column C	Column D
Device	Watts	Hours/day	Watt/hours/day
Fluro Lights (4)	18	2	144
LED reading lights (3)	2	1	6.0
Radio	15	1	15
TV (LCD 26 inch)	100	2	200
DVD	20	1	20
Kitchen appliances	500	25	125
Laptop computer	36	1	36
Water pumps (two by 12 V)	48	0.5	24
Heater pump (winter only)	20	6	120
Total of Column D			690
Plus 15% for system losses			103
TOTAL			**793**

Workings for above	
Total daily energy demand (including 15% excess) is 875 Wh/day =	73 Ah/day
Average solar input 2.0 PSH (6 X 120 watt modules 1350 Wh/day) =	112 Ah/day
Batteries (two sets, each of two 120 Ah)	480 Ah

Chapter 16

Extra-low voltage wiring

Extra-low voltage is the legal and technical term for clean direct current (dc) of less than 120 volts, and for alternating current (ac) of less than 50 volts. Confusingly, the term 'Low voltage' actually extends from Extra-low voltage to 1500 volts dc and 1000 volts ac. Chapter 17 also refers.

Given some dexterity with tools, installing an RV or cabin's typical 12 and 24 volt dc wiring and associated bits and pieces is not that hard to do. You may legally do this with 'non-chassis' wiring and components but front and rear lighting, indicators, electric brakes and brake lighting etc must be done by an auto-electrician licensed to do so.

In all states and jurisdictions in Australia except Victoria, all Low voltage work on Electrical Installations must be done by a licensed electrician and an Electrical Compliance Certificate supplied.

The Victorian exception

For reasons that at best are unclear, Victoria's electricity regulators state that they do not classify RVs as Electrical Installations (but seemingly as 'Appliances'). They are required to meet relevant Australian Standards - and for the first model of any new product range to be inspected and certified. Thereafter, 230 volt wiring and all else 230 volt involved may legally be done by unlicensed workers. No routine inspection or subsequent electrical certification required. The subsequently lower labour cost is almost certainly why so many Australian RVs are made in that state (but also and increasingly in Queensland where normal electrical regulations apply).

Buyers need to be aware that if a Victorian-made RV is sold interstate (particularly in Queensland) , the buyer is likely to need to have the RV's electrics inspected and certified.

RV Books does not suggest nor imply that others are likewise, but has first-hand experiences of two up-market Victorian-made camper trailers with installation faults.

One had so many examples of reverse polarity and other failings, the unit was totally rewired.

Another had the 12 volt supply cable to the slide-out kitchen run (unprotected) such that it was pulled across a jagged metal slideout section each time the kitchen was used. It was eventually cut and short-circuited by that jagged metal (Figure 1.16). Fortunately the owner had protected the trailer's electrics by adding a circuit breaker (that cut the power before the resultant sparks ignited the unit).

Safety

When working with batteries wear fully-covering clothes, protective gloves and a face mask, or at least shatter-proof goggles.

There is some risk of electric shock as any current over 10 milliamps passing through a human body will produce a severe shock and only a little more can be lethal. This is improbable below about 50 volts (a disconnected 12 volt solar module facing the sun produces 17-18 volts) but you need to be careful around even (nominally) 36 volt solar systems: their output can exceed 60 volts and the mild shock could well result in your falling off a ladder.

Always disconnect batteries before doing any work on or around an electrical system. Disconnect the negative (earth) lead first as accidentally touching earthed metal with the spanner then does no harm - but doing that whilst undoing the positive lead (with the earthed lead in place) can lead to that spanner literally vapourising. When reconnecting, and for the same reasons, attach the positive lead first.

The major risk with installed 12/24 volt wiring is excess current heating a cable sufficiently to melt its insulation and then setting fire to itself and nearby material. This is usually caused by a short circuit within an appliance, or by live cables touching. Guard against this by including a heavy fuse, or preferably a circuit breaker, as close to the battery as possible, and firmly securing all cabling against accidental movement. This is discussed in greater depth in the next chapter.

Absolutes

When wiring 12/24 volt electrical stuff there are a few absolutes. Most have to do with not losing energy through so-called 'voltage drop'. Voltage is akin to pressure forcing water through a narrow pipe, i.e. voltage 'pushes' electrical energy through a cable. Even the best and thickest conductor (the copper part of a cable) resists this happening and in resisting the flow, the cable's copper conductor heats up. Generating heat this way is useful if you are designing an electric kettle, but in wiring it is to be avoided.

Current is akin to the amount of water flowing in a pipe. The smaller and longer the cable and the greater the amount of current that it carries, the more the cable heats up and the greater the energy thereby lost. For appliances to work well, this loss should not exceed 1.5%-3.0% (that is 0.17-0.36 volts in 12 volt circuits) between the energy source and whatever is connected to it.

A 3% loss may not seem much: it is only 0.36 volt in 12 volts. The problem however is that most batteries (except LiFePO4s) and appliances are only effective from about 12.6 volts down to about 11.8 volts. If 0.36 volt is lost before it gets to the appliance, that's not a 3% loss of available voltage - it is close to 30%. This tends to be overlooked. LiFEPO4's rarely fall below 12.9 volts in RV usage but there's still no point in losing even minor voltage to save a dollar or three on copper wire.

Fridges are particularly affected by voltage drop: with these 0.3 volts is too high, 0.2 volt is acceptable, and 0.15 volt is a good target.

In practice, one rarely sees an RV fridge that is connected by adequate sized cable. This is commonly due to commercial installers using light cable to reduce their outlay. As a result fridge cabling may have as high as 0.5 volt drop and the fridges can never deliver their intended performance. Further, the 1.5-2.0 metre cables supplied with fridges may themselves have 0.2 or more volts drop. Rewire using at least 6.0 mm cable - it may seem overkill but can transform compressor fridge cooling.

Excess voltage drop may be caused by a DIY installer misunderstanding so-called 'auto cable' ratings (this is covered later in this chapter). It is also often due to following cable chart recommendations that are not applicable to RVs, and is compounded by the lack of any RV industry wiring standard/s. The approach explained later in this Chapter circumvents these problems.

Voltage drop is proportional to cable length: the longer a cable, the thicker it must be. Heavy cable is not cheap. Where possible locate bits and pieces (consistent with ease of access) so as to keep the following cables short: solar modules to solar regulator, solar regulator to batteries, batteries to electric fridge, batteries to inverter and inverter to microwave oven. If it is not feasible to keep a cable short, all is still fine as long as adequately sized cable is used. It just costs more money.

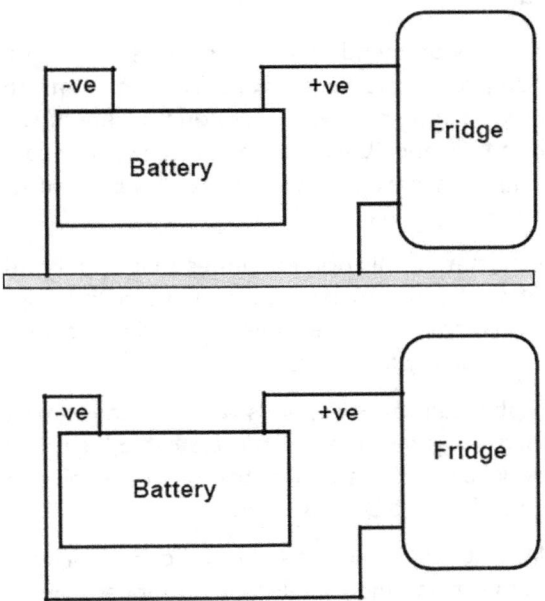

Figure 1.16. Above: earth return.
Below: twin conductor (recommended). Pic: rvbooks.com.au

Electrical circuits that supply dc energy have two conductors: one is positive, the other is negative. Both are identical except for colour.

Some motorhomes (but rarely caravans) utilise the vehicle's chassis as the negative conductor. This saves cable, and because the chassis is so massive, voltage drop across it is reduced. In theory this seems good practice and is recommended by some major equipment suppliers. In practice however, it only works well when done to fully professional standards - but it rarely is. Figure 2.16 show both methods.

Unless the connections to the chassis are extremely well done (ideally by using welded-on studs that act as terminals), such connections invariably corrode and cause problems over time. The practice can also cause electrolysis corrosion through stray current travelling via the radiator and various other paths.

Be aware that most post-2014 motor vehicles use the main earth strap from the chassis to the alternator and/or starter motor battery as an essential voltage reference. If alternator charging is used all negative returns must be taken to the main chassis power post - not directly to the vehicle's

Wire tables

Wire tables show recommended minimum sizes for lengths and current flow, but invariably for single conductors. This may confuse non-electricians, as, with twin conductor wiring, conductor length is twice the cable length. If the distance between the battery and whatever is connected to it is three metres, there will thus be a total of six metres of conductor.

Unless otherwise stated, this book assumes that twin conductor wiring is used. Presenting the information this way is also fail-safe. The worst that may happen is that you'll use cable bigger than needed. But that's a 'good' and not too costly a mistake.

The current a cable can carry depends on its conductor size. The International Standards Organisation (ISO) rates cable logically by quoting the cross sectional area of the conductor in square millimetres, usually abbreviated to mm², e.g. ISO 35 is 35 mm².

Multi-strand ISO cable suitable for solar and other wiring is sold by electrical wholesalers and marine electrical suppliers, but only rarely by auto parts stores - see 'Auto cable' (below).

Cable made for fixed domestic and other wiring is ISO rated, but is best avoided for 12 volt RV wiring because it has only seven strands and lacks flexibility. It may fracture over time. That made for flexible power cords etc has many finer strands and is more suitable for RV use.

The cable most commonly sold in the USA (and some other areas) uses the American Wire Gauge (AWG) rating system. The lower the AWG number the thicker the conductor. The B&S (Brown & Sharpe) rating system is identical - except for minor differences in a couple of sizes.

Most conversion charts 'round-off' sizes to the next higher (i.e. thinner) even-number AWG/B&S equivalent, but this increases voltage drop. Table 1.16 shows the recommended thicker equivalent.

ISO (mm²)	0.75	1.0	1.5	2.5	4.0	6.0	10	16	25	35	50	70	95	120	150
Auto cable	2.5	3.0	4.0	5.0	6.0		8.0								5/0
AWG	18	17	15	14	12	10	8	6	4	2	1	2/0	3/0	4/0	5/0
B&S	18	17	15	14	12	10	8	6	3	2	0	2/0	3/0	4/0	5/0

Table 1.16. Approximate relationship of the four most common cable 'standards'. Where discrepancies arise, most tables like this round down to the next thinnest size: but that increases voltage drop. This table thus rounds up to the next larger size. Table: rvbooks.com.au. This table may be reproduced subject to full acknowledgment.

If money is not too tight, the very best choice is the so-called 'tinned copper' ISO-rated cable sold mainly by specialist marine electrical stores and also by Springer Low Voltage Electrics via its stores in southern Queensland.

Auto cable - a very real trap

Outside the USA, almost all appliance manufacturers specify ISO-rated cable. To them, a 4.0 mm cable implies a 4.0 mm² conductor. This introduces a trap for the electrically unwary.

Many countries, including Australia, use local and Asian 'auto cable' for extra-low voltage wiring. Auto cable is that stocked by auto parts and hardware stores. For reasons that defy sanity, auto cable is rated and marketed by its overall diameter (i.e. including its insulation). But as insulation thickness varies hugely from brand to brand, and even types of the same brand, the resultant 'rating' shows only the size hole you can just poke the cable through.

Thus '3.0 mm' auto cable can (and often does) have a conductor area of only 0.5 mm² to 1.5 mm². The commonly-used '4.0 mm' auto cable can be as small as 1.25 mm². That sold in Australia and New Zealand is more likely to be 1.8-2.0 mm² but only rarely larger.

The '6.0 mm' auto cable has less variation. It is likely to be 4.4-4.8 mm². The 8.0 mm size (similar to AWG/B&S 8) varies from about 7.0 mm² to a little over 8.0 mm².

Auto cable is thus a real trap and catches out any number of people. There is nothing wrong with it as such: it is readily available and affordable. If specified correctly it works just fine. But you must know its conductor area in mm². This is often printed in small type on the side of the cable drum. It is always revealed in the manufacturers' technical data and most stores have that. Insist on seeing that data because making a mistake here can result in a system that barely works.

Current rating

A cable's 'current rating' can be even more misleading. It is simply a fire rating that indicates the current a cable can carry before its insulation begins to melt, and that varies from brand to brand.

Few auto parts suppliers, let alone buyers, understand 'current rating.' Customers often ask for cable to carry (say) 30 amps but not (and vitally) specifying its length. They are sold a '30 amp' fire rated cable regardless. Ignore this rating. Select cable *only as shown in the next few pages of this book* and you will automatically be well within its fire rating and assured acceptable voltage drop.

A huge number of cabin and RV electrical failings are directly due to excess voltage drop, especially those relating to fridges. Cabins are more likely to be wired up by licensed general electricians but not all know about this auto-cable rating anomaly. Most auto-electricians are aware of it however, and not least because it is covered in Caravan & Motorhome Electrics as well as here - and both double as auto-electrical text books. The author also supplied the regular technical column in the auto-electricians' trade publication, AEAN, from 2007 to 2015.

Specifying cables

Most books and magazine articles in this area include complex tables that recommend cable size for various distances and current flow. Almost all assume a voltage drop of 5%, but that is too high for cabin and RV cabling. Further, most show only AWG/B&S sizes, with no guidance as to what to do if that cable is not available.

Since 2007 this and its associated books suggest a different approach to voltage drop calculation by adapting a very minor variant of the ISO (International Standards Organisation) standard constant associated with voltage drop.

ISO Standard re voltage drop

The exact ISO 'voltage drop constant' is 0.0164. Rounding that up to 0.017 is easier to remember and introduces results that are a tiny increase in recommended cable size but often usefully closer to those commercially available.

Instead of confusing or misleading tables, the method below results in the approximate cable size (in square millimetres) for any desired combination of length, current and voltage drop. Pick the cable nearest to that size (but *never* smaller).

Where:

L is total conductor length (in metres)

I is total current (in amps)

Vd is desired voltage drop (in volts)

L X I X 0.017 divided by **Vd** = conductor size (in mm²)

Example 1: a 12 volt 1500 watt inverter drawing 130 amps, connected by twin cable to a battery one and a half metres away (i.e. three metres of conductor). We wish to keep voltage drop to 0.2 volt.

Applying the formula we thus have 3.0 X 130 X .017 - divided by 0.2. The result is 33.15 mm². The comparison table 1.16 shows the closest size to be 35 mm² (also 2 AWG and 2 B&S). The sum works equally well for (say) 24 volts.

Example 2: a 1500 watt (but 24 volt) inverter draws 65 amps and is connected by twin cable at one and a half metres distance from the battery. The same percentage* drop at 24 V is 0.4 volt (not 0.2 volt).

The sum is thus 3.0 X 65 X 0.017 - divided by 0.4.

The result is = 8.287 mm². That's an easy one too - about 8 mm auto cable, 8 AWG and 8 B&S. The closest ISO size is 10 mm² and far from overkill.

* That the required cable is only one quarter the size (not half) for 24 volts often surprises. The explanation is that for the same watts, the current draw at 24 volts is halved and, as it is primarily the *percentage* volts drop that matters, at 24 volts that's 0.4 volt - not 0.2 volt.

12/24 volts - the choice

The above shows the advantage of using 24 volts, but little 24 volt equipment is now available for RVs excepting lights, water pumps, chargers and inverters. For a vehicle with a 24 volt alternator, consider retaining the 24 volt starter battery, then add a 24/12 volt dc-dc charger and 12 volt battery bank. Or stay with 24 volts, and run as much as possible from 230 volts via a 24 volt to 230 volt inverter. See also Chapter 1.

Fixing inadequate wiring

If existing cabling is too light in a charging or a fridge circuit, replace the cable, or run a parallel cable to share the load.

In lighting circuits, replacing inefficient lights by items that draw less current will reduce or eliminate this problem. Replacing 20 watt incandescent globes by 10 watt halogen globes in the 1990s resulted in similar light levels for half the current - and hence half the voltage drop. The now universally accepted white and warm-white LED lights draw so little current for more than adequate light that 1.5 mm² cable can be virtually relied upon to cope. Anything smaller is best avoided, but for reasons relating to being to easily broken rather than current carrying capacity.

Which cable is which - colour coding

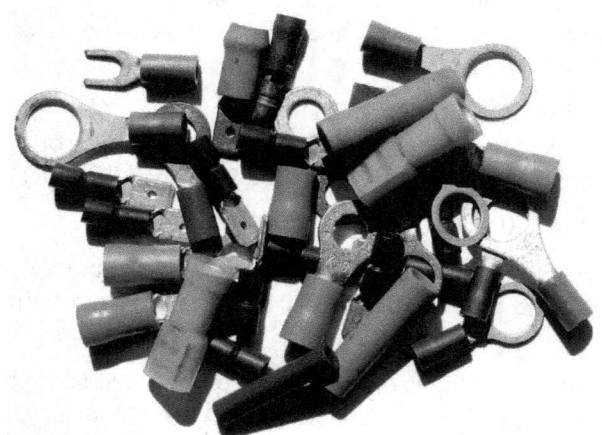

Figure 2.16. Crimp lugs vary in quality - from dreadful to aircraft certification. The best have seamless construction, not folded. They stocked by electrical wholesalers (they will supply retail if you pay using cash).

Any colours tend to be used for extra-low voltage wiring but it is common to use red for positive, and black (or in some countries yellow) for negative. But don't rely on this. Jayco for example, uses black for positive and white for negative. Engel and many others use black with a white stripe for positive and plain black for negative.

Solar modules, batteries, appliances, wiring diagrams etc., indicate positive by a plus (+) and negative by a minus (-). They may alternatively be marked +ve and -ve respectively. Earth connections are marked either 'earth', or in the USA, 'ground'. Positive leads connect to positive (+) terminals, negative leads to negative (-) terminals.

Some electrical wholesalers supply numbered or lettered tags for cable identification. This is worthwhile for subsequent fault finding.

Making connections

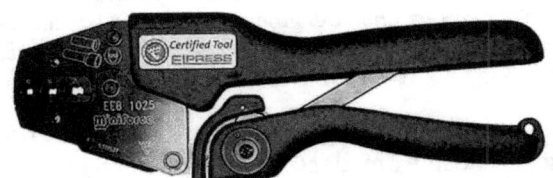

Figure 3.16. Form crimps only by using a high-quality tool. Do not attempt to do this with pliers. Pic: Digi-Key.

Cables connect to electrical devices via plugs and sockets, crimp lugs or are secured by set screws. Crimp lugs work by squashing wires tightly inside strong metal sleeves. Crimping works well as long as you use high-quality lugs, but cheap ones tend to corrode and/or work loose after some years.

These lugs are colour coded. Red is used for 1.5 mm^2 cable, blue for 1.8-2.5 mm^2, yellow for 4-6 mm^2. It is essential to use the right size lugs and a good, purpose-made crimping tool.

If used correctly, these tools create a cold weld. Pliers only *appear* to work. They cannot form that desirable 'cold weld'. Connectors crimped that way may seem to be just fine, but fail due also to corrosion after a year or two.

Crimp lugs suffer from a wider range of quality than is normal with electrical products. The cheapest are made from folded-up sheet metal, you can see the join if you look inside the crimp lug. It is well worth buying aircraft-quality lugs. These lugs are formed from extruded tubing, they cost a lot more but their use will assist to avoid subsequent problems.

Soldering crimp lugs is absolutely not advisable, especially for wiring that is subject to vibration: it locally stiffens the copper conductor which may cause it to fracture over time. It may also cause the copper to corrode.

Connections to solar modules are made by crimp lugs, or by forming the copper into a loop held by a screw and shaped washer supplied with the module. Battery and inverter cables need crimping hydraulically (manual crimping tools are far too small) have these done by an auto-electrician.

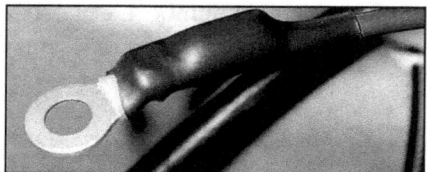

Figure 4.16. Protect and strengthen the crimped connections by using (heat) shrink-wrapped tubing - both from electrical wholesalers.

Support cables against movement by using plastic ties at not more than 400 mm spacing. Those from electrical wholesalers are usually much stronger than those from hardware stores.

Protect cables also against abrasion, but don't run them inside closed tubing - that presents major access problems later. Instead, wrap cable/s in plastic spiral binding.

An exception is where cable must run where later access is not feasible. There, use rigid or flexible electrical conduit - having it slightly oversize assists pulling cables through it. Use a vacuum cleaner to suck through a cord - then use that cord to pull through the cable. If the cable sticks, a squirt of WD-40 or French chalk works wonders. Neither damages the cable.

Joining cables

Figure 5.16. This waterproof connector accepts six cables.

Multiple cables may be joined or terminated via connector strips. They are made in sizes to accommodate three to twenty or more cables.

Some accept variously different sizes (of up to 100 mm² or so). They are also made as small lidded-boxes in a choice of red, black or clear to indicate polarity and earth.

Heavy cables are best joined via terminal posts, not the unsightly and dangerous cluster of leads often seen hanging off battery terminals. The latter is particularly poor practice with vented lead-acid batteries because nearby cable connectors and cable become corroded by acid fumes.

Fuses

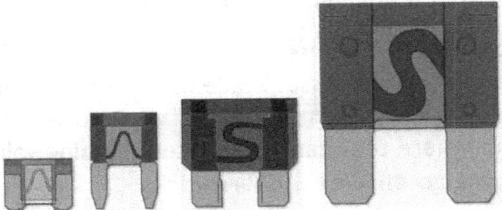

Figure 6.16. Blade fuses. Use the larger size, for 20 amps and over. Pic: grainews

One of the more serious electrical faults involves live conductors accidentally touching. Such heavy current may then flow that the cable melts and ignites nearby flammable material.

Circuit breakers or fusible links protect against this. They are inserted in the battery end of cables to protect the cable. Fuses are also used to protect appliances against fire or damage if a fault causes them to draw excess current.

These fuses need to be located close to the appliance. If you add a fuse, do it properly. Many electrical faults, particularly with big fridges, are caused by the fuse holders themselves burning out.

Avoid tubular glass fuses and their holders: they are seriously unreliable. Use the blade type instead. These fuses are made in different physical sizes. Use 'maxi' (industry term) size fuses and holders for 20 amps and over. Smaller sizes can overheat and even melt.

Figure 7.16. Typical 100 amp fusible link. Pic: Ford Motors.

Above 30 amps it is better to use so-called fusible links (Figure 8.16). Fusible links are high current fuses made to higher standards. They are like oversized fuses but have screw or similar positively locking connectors. Locate them where they cannot splatter anything valuable with molten lead - some go off like $100 fireworks.

Circuit breakers

There have been major developments in fuse technology, but following a major research study, insurers found a statistically significantly lower incidence of electrical fires where circuit breakers are used.

As a result, the use of circuit breakers (rather than fuses) to protect cables is strongly recommended in electrical codes. It makes sense to follow this recommendation.

Some circuit breakers also double as main circuit switches. They are invaluable for cutting power manually if a fault develops that does not trip the circuit breaker (such as a partial short circuit in inaccessible wiring). Such ability is much faster and safer than trying to replace a fusible link.

In deciding whether to choose circuit breakers or fuses, bear in mind that high-quality circuit breakers, such as that shown in Figure 9.16, are not cheap. Those that are cheap tend to be unreliable and temperature dependent.

*Figure 8.16. DC 12/24 volt 60 amp resettable circuit breaker.
Pic: Asdomo.*

Unless you are prepared to pay for high-quality circuit breakers, settle for 'fusible links' (from auto electrical stores). Safety-wise, they are a better proposition than cheap circuit breakers.

Master switches

Unless protected also by a circuit breaker or fuse, never install a master switch in any high-current circuit carrying dc. If there is a short in that cable, opening an unprotected switch whilst it is carrying excessively high dc current will cause the opening contacts to form an arc. This ionises the air - enabling the arc to extend and be maintained. Where currents are high enough a dc arc can almost instantly vapourise a switch thus spraying molten metal. This effect does not happen with ac because both voltage and current pass through a zero current state at 100 times a second. This extinguishes any arc.

Clipsal advises that its 230 volt ac circuit breakers can be used for dc applications below 48 volts but warns that they take longer to trip. It seems pointless to use the ac version, however, for 12 and 24 volt dc use as the dc circuit breaker version costs only a little more.

Switches

Domestic 230 volt switches are not intended for direct current (dc) operation, but they work reliably enough if used at no more than 20% of their (typically 10 amp) rating.

Amps times volts equals watts so a 20 watt, 12 volt load draws 1.66 amps. This current is readily handled by a 10 amp 230 volt ac switch.

Two 10 watt, 12 volt loads equally draw 1.66 amps and are thus also switchable this way.

Alternatively, small ac/dc rated toggle switches are available from component suppliers such as Altronic and Jaycar Electronics. Most of these are rated at about 10 amps ac and 2 amps dc.

Twelve volt plugs and sockets

Early editions of this book (and my companion books in this field) advised users not to consider cigarette lighter plugs and sockets for anything drawing more than an amp or two - and preferably not at all. This was because, until recently, most lacked mechanical locking. When they inevitably worked loose they arced internally, becoming an only too-real fire hazard.

Figure 9.16. Hella 16 amp locking plug. Pic: Hella

Whilst this advice generally still stands, Hella's 8 and 16 amp version of these plugs and sockets include effective mechanical locking. Engel too provides locking ones - with some having an internally fused plug.

The (UK) Bulgin company make excellent marine-rated 12 volt plugs and sockets. Another alternative is the very much larger two-pin plugs and sockets available from Anderson, Camec, Clipsal and others. Avoid eBay specials and the many Anderson look-alikes. The fire risk is too high to attempt to economise.

Chapter 17

Low voltage wiring

Low Voltage is defined (in Australia and New Zealand) as that exceeding 50 volts ac and not exceeding 1000 volts ac. Our so-called mains-voltage used to be defined as 240 volts but to accord with EU standards (and mainly to enable local manufacturers to sell product overseas) was, in 2000, legally defined as 230 volts +10%/-6%. In practice much of it is still about 240 volts.

The 12/24 volts associated, for example, with cars and RV batteries, is legally known as Extra-low voltage. It is defined as being below 50 volts ac, and 120 volts (ripple-free) dc.

The installation of Low Voltage (i.e. mains-voltage) wiring and associated work in those defined legally as 'electrical installations' may only be done (in Australia) by licensed electrical contractors who are responsible for certifying that the work has been done correctly. This restriction applies if the installed wiring is supplied only by an inverter or generator (if they are connected to fixed wiring) i.e, even if there is no external mains connection.

For reasons that seemingly defy sanity however, Victorian electrical authorities do not classify RVs as Electrical Installations, but rather as 'Electrical Appliances'. They are required to meet Australian standards, but the 230 volt wiring and all else (230 volt) involved may legally be done by unlicensed workers: nor is subsequent routine inspection or certification required. This is almost certainly why most Australian RVs are made in Victoria.

RV Books does not suggest that such RVs are not wired correctly, but has first-hand evidence of one that demonstrably was not. Buyers need to be aware that an Electrical Certificate is likely to be required if re-registering an RV in another state (particularly Queensland).

This Chapter is thus intended only as a description of that legally required, and may provide an indication if any does not. It is not however intended as inducing those lacking electrical certification to attempt work of any kind (beyond changing fuses and light globes) to work on Low Voltage systems.

Experience shows that some electricians seem unaware that RV electrical systems are substantially different from domestic practice, or do not always following the legal requirements. Caravans, campervans, motorhomes and some cabins are exposed to potentially dangerous conditions that are less likely in fixed installations. The main risks are increasingly recognised and many requirements have changed in recent years. Some such requirements, such as that for double-pole switching for RVs, and the neutral-earth connection, that for RVs relies on that at the source of the electricity supply (i.e. not within the RV) are substantially different from domestic practice. This is covered in depth later in this Chapter under the sub-heading 'Vital Exceptions'.

Retired electricians reading this are respectfully reminded that much of the Wiring Rule requirements they knew so well were substantially changed in 2001 and 2002, and again in 2007 and 2008. They were then substantially Amended in 2012 with major changes to the requirements for inverters and generators in RV systems. A new version was published in 2018.

If seeking electrical work to be done on an RV it is advisable to ask a few discrete questions to ensure the electrician really is aware of the current requirements (precised below) - and then check they are followed.

Relocatable premises

In both Australia and New Zealand 'relocatable premises' must meet the requirements of requirements of AS/NZS 3000:2018, plus those of AS/NZS 3001:2008 (as Amended in 2012).

Relocatable premises include registrable vehicles offering accommodation, e.g. caravans, camper vans, motorhomes, camper trailers, and livestock and other transporters that have accommodation included. In New Zealand this is extended to include many other categories including vending vans, mobile classrooms etc. Also included are non-registerable premises such as re-locatable homes, transportable huts (that includes cabins), and rigid and non-rigid annexes to the above. There are several further categories (such as canteens, mobile toilet blocks) but these are outside the scope of this book.

Permanent connection exceptions

The provisions of AS/NZS 3000:2018 (only) apply to re-locatable premises where an electricity supply has been permanently connected and where such premises are not intended to be relocated, e.g. coaches or caravans that have been rendered immobile, and cabins that are permanent structures. For these, the requirements of AS/NZS 3001:2008 (as Amended in 2012) do not apply.

Vital exceptions

As well as many minor differences, there are a few major and vital differences between electrical installations supplied by a connecting cable from an external socket outlet (e.g. in a caravan park or home) and those that are permanently connected to an electricity supply (as in a house or business).

One difference particularly relates to earthing. In some countries (including Australia and New Zealand) the neutral line is connected to earth in various specified locations. With this so-called MEN (Multiple Earth Neutral) wiring, houses and business premises have an earth-neutral link made within those premises.

In re-locatable premises, such as caravans and motorhomes that are supplied by a connecting cable from an external socket outlet, that earth and neutral link is at the supply side of the connecting cable and site outlet, i.e. not within the re-locatable premises. To do otherwise would require the caravan owner to ram a copper earthing spike deeply into the ground every time the caravan was moved - and impossible on concrete and other hard or non-electrically conductive surfaces for the electricity supply to be used.

Whilst much of the above Standards are common to Australia and New Zealand, there are some differences between their respective requirements. These include an Australian (only) requirement for circuit-breakers and switches to be double-pole (i.e. to switch both active and neutral leads).

In New Zealand, however, whilst the location of the earth-neutral (MEN) connection is now mostly as above, many older NZ caravans and motorhomes still have earth and neutral linked within the vehicle. That earth/neutral link must be retained if no RCD (residual current device) has been installed.

Some NZ RVs have permanently connected supply cords that have an inbuilt RCD. With these, the earth/neutral link must not be made within that RV.

Power into the vehicle (socket outlets)

Cable rating	Conductor area	Length
10 amp	1.0 mm²	25 m
10 amp	2.5 mm²	60 m
10 amp	4.0 mm²	100 m
15/16 amp	1.5 mm²	25 m
15/16 amp	2.5 mm²	40 m
15/16 amp	4.0 mm²	65 m

Table 1.17. Connecting cable lengths and minimum conductor sizes generally applicable to RVs.
(This is actually Table 5.1 of AS/NZS 3001:2008 (as Amended in 2012. It supersedes that in previous editions of this book. Data: Standards Australia).

Caravan parks are legally required to have:

• Socket-outlets complying with AS/NZS 3112 (3 flat pins) and rated at not less than 15 amps.

• Or, (in Australia) socket-outlets complying with AS/NZS 3123 (3 round pins) and rated at not less than 20 amps.

• Or, (in New Zealand) socket-outlets complying with IEC 60309-2 (round pin) and rated at not less than 16 amps or 32 amps.

Each socket-outlet has to be protected by a circuit breaker and an RCD (residual current device).

Supply cables

From 2008 onward a change in the Standard ensured a maximum 5% uniform impedance i.e. resistance to alternating current. This limit enables sufficient current to flow for circuit breakers to open within the 0.4 second in time to save human life. This enables a wide choice of supply cable lengths and conductor sizes. That current flow is restricted if supply cables are joined end to end. The Standard thus now requires supply cables to be of one single unbroken length. This should be explained to the many RV owners that still interconnect cables unaware of the reason why it is prohibited.

It is advisable to use heavier cable than that specified if there is any risk of it being damaged.

10-15 amp adaptors

The use of 10-15 amp adaptor leads, or filing down 15 amp plug earth pins for use where there are only 10 amp outlets has always been potentially dangerous. Doing so is now also illegal.

The only practicable and legally acceptable solutions (for direct home connection) is to have a licensed electrician install a 15 amp circuit and socket outlet.

For use where 10 amp outlets are only occasionally available is to use a specially made 15 amp to 10 amp adaptor cable that incorporates a 10 amp circuit breaker. It is available under the trade name Ampfibian (Figure 1.17).

Figure 1.17. Ampfibian 15 amp to 10 amp adaptor. Pic: Ampfibian.

Where the RV does not need 15 amps (i.e. where 10 amps will suffice) another solution is to change the RV's inlet plug and circuit breaker/s changed to 10 amp and use a 10 amp cable. A direct replacement 10 amp RV socket inlet is now available (from Camec).

Another solution is to omit the socket inlet at the RV end and have the supply cable permanently attached. Some restrictions apply regarding a waterproof enclosure and mechanical fastening for the stored cable.

Protecting against 'dirty' power

Some (mainly outback) caravan parks have generators that produce 'electrically dirty' power. The more basic 230 volt generators do this too. To protect against this, Ampfibian has (mid-2018) produced a 'power protection unit that may legally be plugged in-line with its 10-15 amp adaptor. This unit is however a 'once-only' device. It protects your computer or whatever, but is destroyed whilst so doing.

Vehicle inlet socket

The RV standard socket inlet must comply with AS/NZS 3109.1, AS/NZS 3123 or (for NZ), IEC 60309-2. It must be mounted so that the earth pin is not uppermost and, if on the same side as an entry door, not less than 1.5 metres from any entry door, and not less than 150 mm from any opening.

Polarity and double pole switches

A possible hazard for cabins and (pre-2000 or self-made) RVs is that the incoming power may be incorrectly polarised (i.e. active and neutral leads reversed). With correct polarity, a switch that breaks the active lead only ensures a connected appliance is electrically dead. But, if the polarity is reversed, that switch breaks the neutral line only. The still live active line remains connected to a light or appliance presumed as 'turned off'. This can be fatal if someone, assuming power is off, attempting to replace a light globe, accidentally breaks the glass and contacts the still live bits now exposed.

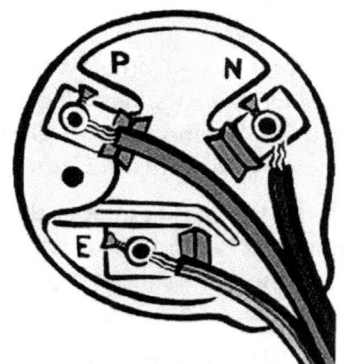

Figure 2.17. P (Active) red or brown. E (Earth) green or yellow & green).

A possible hazard for cabins and (pre-2000 or self-made) RVs is that the incoming power may be incorrectly polarised (i.e. active and neutral leads reversed). With correct polarity, a switch that breaks the active lead only ensures a connected appliance is electrically dead. But, if the polarity is reversed, that switch breaks the neutral line only. The still live active line remains connected to a light or appliance presumed as 'turned off'. This can be fatal if someone, assuming power is off, attempting to replace a light globe, accidentally breaks the glass and contacts the still live bits now exposed.

Reversed polarity is mostly caused by people (illegally) making up their own supply cables and reversing active and neutral at one or other end of the cable. It can also occur through incorrectly wired inlet and outlet sockets etc. even in new RVs.

To safeguard against this, all new Australian RVs and re-locatable premises have since 2000 or so, been required to be protected by double-pole switches: i.e. that cut both active and neutral lines.

Whilst a few pre-2000 RVs have been electrically upgraded, incorrect polarity is still an ongoing problem. Many RVs still have only single-pole switches. So do any number installed subsequently by people (allegedly including a few manufacturers) unaware of the requirements - or the need for them.

Whilst not obligatory it makes sense to also use double pole switches in cabins - particularly if moved from time to time.

For correct polarity, the active (brown or red) and neutral (blue or black) conductors must be connected to the corresponding terminals of plugs and sockets. Red or brown connects to the active (A) terminal, black or blue to the neutral (N) terminal, and green or green/yellow to the earth terminal - Figure 2.17.

Because New Zealand has tighter control over supply outlet and cable polarity, plus obligatory inspections, single-pole contact breakers and switches are deemed acceptable there.

Polarity testing

Incorrect polarity can readily be checked by using a plug-in tester (Figure 3.17). These devices typically have three lights: red, green and amber, that light up in various combinations to indicate correct or incorrect polarity and also the presence or otherwise of an earth connection.

Figure 3.17. Plug in polarity tester. Pic:solarbooks.com

Polarity testers also indicate whether a supply cable or caravan park power outlet is wired safely. Not all such power outlets are, particularly in outback areas where the source is a diesel generator. This is not good, but less of a risk providing your RV has the legally obligatory (in Australia) double-pole switching.

A polarity tester warns of faults but does not indicate their location. A fault could be in the outlet checked, the cable leading to that outlet or further back in the wiring. To find out, start at the furthest point (e.g. caravan park socket outlet) and work forward checking each stage.

It is worth having a polarity tester permanently plugged into a socket outlet inside the vehicle (they cost about $15 and are stocked by hardware and electrical suppliers).

Inverters

As noted in Chapter 8, there are two main types of inverter: non-isolated and isolated. A non-isolated inverter usually has an output that is 115-0-115 volts, and with one side of the output usually connected to one side of the battery input, introducing a possible safety risk. An isolated inverter has its 230 volt output totally free of any connection to the battery. Whilst the risk of using a non-isolated inverter is low, that risk is not worth taking to save a few dollars. RV Books strongly advises to only buy a high quality isolated inverter.

Most such inverters are intended to be free-standing. They have power outlets into which appliances (or a multi-outlet power board) must be plugged directly. These inverters absolutely must not be connected to fixed mains-voltage wiring - even if mains electricity is never used. If in doubt about compliance, show this page to a licensed electrician and seek his or her advice.

Change-over switches

A requirement, where a generator or inverter is connected to 230 volt wiring, is for a 'break before make' double-pole change-over switch. This switch ensures that generator or inverter output cannot be accidentally fed into the grid network. It is intended to protect electricity workers working on the network (with power cut off from the main grid) and thus assumed safe for handling. Here again, this is licensed electrician territory.

Cabling

Low voltage lighting cabling is usually 1.5 mm² (but see below). Power cabling is usually 2.5 mm². If Low voltage wiring is covered by thermal insulation, 2.5 mm² cable must be used also for lighting circuits. The so-called 'building cable' (that typically has about seven strands of larger size) is liable to mechanical failure under vibration. It is fine for permanent installation in cabins, but not in RVs.

Separation of Low and Extra-low voltage wiring

Low voltage and Extra-low voltage wiring must be physically separated, or the Extra-low voltage circuit must use Low-voltage insulated cable. The bulk of such cable may however make the latter impracticable. It is simpler, and safer to install Low voltage and Extra-low voltage cables well apart.

Wiring protection

Low voltage wiring must be protected against damage possibly caused by virtue of its location, or by additional protection such as rigid or flexible conduit. It must be protected by rubber or plastic insulating grommets where it passes through metal. Where wiring passes through timber it is good practice to smooth off the edges of the holes.

All such wiring must be arranged so as to limit movement or stress on conductors at points of termination. Unenclosed cables must be securely fixed at intervals of not more than 300 mm by suitable clips, saddles or clamps; or securely fixed at intervals of not more than 500 mm when in an enclosed space, and supported on a horizontal surface such as a ceiling or frame member. No further fixing is required where wiring passes through bushings or grommets in ribs in an enclosed space and such ribs are not more than 500 mm apart. All is set out in

AS/NZS 3001:2008 as Amended 2012 (sections 3.4.2 and 3.4.3). A currently licensed electrician should be able to advise.

Kitchens and bathrooms

There are any number of regulations governing the location of light fittings and socket outlets in or near kitchens, bathrooms, showers and water. These regulations are laid down in the Standards and are well known and understood by most electricians, but sometimes (illegally) ignored by RV builders.

Certification

The previous need for independent inspection changed around 2002. Now, (in states and jurisdictions other than Victoria) the licensed contractor or manufacturer takes responsibility for RV systems and issues a certificate accordingly.

Updating installations

In 2000, electrical requirements called for RCDs to be obligatory for all new work, resulted in the death rate in Australia from electrocution (mostly from supply cable faults) halving to now under fifty a year.

Electricity regulations specify only *minimum* requirements. Electrical installation is a highly competitive business so these minimum requirements will be met unless otherwise requested. There is a good case for exceeding these requirements. For example it is legal to run power and lighting from the same circuit, but it is better practice have them separated, each protected by an individual CB/RCD.

Chapter 18

Installing batteries

Large capacity batteries are heavy and awkward. Two or more smaller ones are easier to handle than one big one. For 12 volt systems, you can have 12 volt batteries connected in parallel, or you can have one or more pairs of two 6 volt batteries connected in series.

Don't buy into arguments for or against series or parallel connections: the pros and cons largely cancel out. Either way works fine subject to the following.

Parallel connection

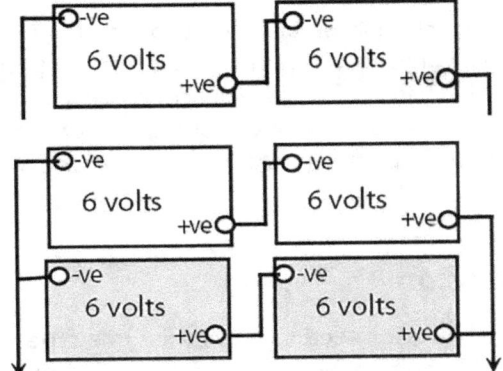

Figure 1.18. Upper: twelve volts can be obtained by connecting to 6 volt batteries in series.
Lower: to obtain higher capacity (here at 12 volts) connect additional pairs of batteries in parallel. Pic: rvbooks.com.au

Parallel-connected batteries must be of the same nominal voltage, and of generally similar condition. Contrary to campfire mythology however, it is fine to parallel batteries of even widely different amp hour capacities. Each absorbs the charge current it needs and each delivers its proportional share of load current.

Any combination of the same batteries *always* provides the same amount of energy - there is no way you can get more energy from batteries than the energy that is already in them.

Battery companies generally advise limiting the number of parallel connected batteries, or strings of series/parallel batteries. Most suggesting limiting it to four batteries or four series-connected strings of batteries, but the giant Exide corporation suggests that ten is still fine given the safeguards outlined above.

Battery makers advise against parallel connecting conventional lead-acid batteries directly with gel cell or AGM batteries. They also advise against charging conventional lead starter batteries and gel cell or AGM auxiliary batteries from the same source. This is because both of the latter have far greater charge acceptance, and would thus starve a lead acid starter battery.

This issue is resolved by including a voltage sensing relay that automatically ensures the conventional battery is adequately charged before further batteries are paralleled for charging. The relay also disconnects the protected battery if the voltage across it falls too low. Details of this form of protection are in Chapter 24.

Many post-2013 vehicles (those built to Euro 3 and 5 emissions standards) have variable voltage alternators. Voltage-sensing relays cannot be used with these alternators. The only successful way to charge auxiliary batteries with such alternators is via a specialised form of dc-dc charging. Please refer to Chapter 6 (and for ongoing updates) rvbooks.com.au.

Series connection

Series-connected batteries need to be of the same type, amp hour capacity and general condition. Voltage is additive. Current output (and charge acceptance) is limited by that of the lowest amp hour capacity battery in the chain.

Battery location and care

With caravans particularly, locate heavy batteries toward the centre of the vehicle and securely held in place. They will be damaged if allowed to rattle around. Consider using a slide out battery holder, or cutting a hatch in the floor to gain access from above.

Ideally, charge the batteries via a close-by dc-dc alternator charger fed from the alternator via adequately heavy cable and (for trailers) via an Anderson plug and socket.

Battery installation and ventilation

Battery vendors' initial claims that sealed lead acid and AGM batteries need no ventilation were rapidly withdrawn. Battery makers insist that ventilation is still essential for all such batteries. Despite this, many RV makers locate sealed batteries in unventilated enclosures, often under the bed. The very batteries installed are likely to carry a warning that ventilation is essential. There are no legally enforceable standards in this area, but there is a legal 'Duty of Care'.

Battery compartments must be ventilated at their top and bottom. Ensure there are no lips that can trap gases at the top of the enclosure. The (Australian) Clean Energy Council suggests that, for batteries up to 350 amp hour or so, a couple of vents or holes, each of 502 mm to 1002 mm, at the very top and very bottom of the enclosure should suffice.

Do not house circuit breakers, fuses, switches or any other electrical device in the battery enclosure. There is no risk in normal operation, but very much so in the event of an electrical fault causing serious overcharging. This is likely to cause a battery pressure vent to open, releasing potentially highly explosive hydrogen and oxygen that, if in a sealed enclosure, can be triggered by a spark.

Under normal conditions, gel, AGM and lithium produce little or no dangerous hydrogen gas. The little gas that escapes is negligible. However, as with all batteries, some heat is generated during charging. To ensure the longest possible lifespan, it is important for this heat to be removed from the battery as quickly as possible. Little information is available about ventilating LiFePO4 batteries (that emit no gasses). The International Fire Code (IFC) and NFPA 1: Fire Code, however, states that 'Li-ion and lithium-metal-polymer batteries shall not require additional ventilation beyond that which would normally be required for human occupancy.'

Battery and cable protection

Some form of safety mechanism must be provided to safeguard batteries from accidental gross discharge. This is also required to prevent their energy burning out (and/or short circuiting) cables connected to them, or were a short circuit to form within a faulty connected appliance. As explained in Chapter 16 this better done by circuit breakers, not fuses.

Ideally one major circuit breaker (that also doubles as a main switch) should be included in the main positive cable. It should be located as close as possible to the auxiliary battery, and readily accessible, but not in the same compartment as that battery.

Separate smaller current circuit breakers are then installed downstream from the main circuit breaker to feed various sub-groups, such as lighting, air conditioning and water pumps.

To save cost some RV makers install a fusible link (Figure 7.16) in place of a main circuit breaker/master switch. These links bolt into place. They are initially cheaper than a main circuit breaker, but need careful location as they may splatter molten lead some distance if they blow.

Both main fuse or main circuit breaker should be rated at the lesser of about 70% of the maximum safe current that can be carried by the cable, or 30% more than the highest current normally drawn through that cable. If in doubt, seek the advice of an auto-electrician.

Figure 2.18. Copper plated brass battery connectors.
Pic: original source unknown.

Never insert a fuse or circuit breaker in the lead from the alternator to the battery. To regulate correctly, the alternator relies on knowing the battery voltage. If this is lost, the alternator voltage may rise instantly to 100 volts or more, burning out the alternator's internal diodes. Make totally sure that this lead is well-insulated and securely located.

Battery cables and connections

Battery cables are best attached by crimping or clamping - not by soldering. Crimps of this size need a hydraulic tool. Have this done by an auto-electrician.

Poor quality battery connectors cause ongoing trouble. If overtightened, they may crack and break in half. The best ones are made from oil-impregnated bronze or copper-plated brass (Figure 2.18). Some have a separate nut and bolt clamp for attaching the cable lug. They are sold by marine electrical suppliers.

Power posts

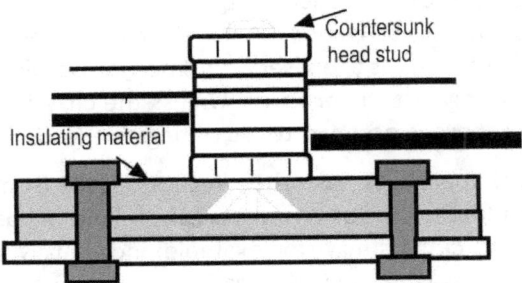

Figure 3.18. This power post was made from scrap lying around a workshop.

Many systems have multiple cables attached to battery terminals - with varying degrees of effectiveness and probability of working loose. Also, with the unsealed batteries still used in some major home and property solar systems, the associated cables and terminals may be attacked by corrosive gases.

Avoid the above problems by taking short heavy positive cable from the battery to a fusible link, or circuit breaker and from there to one or more nearby common power posts, or heavy junction boxes.

Such posts (Figure 3.18) are readily made from scrap material. They are also available commercially.

If using chassis negative returns, have a threaded stud securely attached (ideally welded) to the chassis.

Figure 4.18. Connectors such as this are available in any number of cable sizes and configurations.

Battery terminals and posts need electrical protection. Insulate the posts by plastic or rubber protective boots.

Using an RVs' steel chassis as a common conductor saves cable, but has known adverse effects. If used for a motorhome it can cause serious radiator corrosion if added spotlights have a return circuit via a radiator mounting. Even gearboxes can accidentally become part of a return path (and if that happens can be ruined). Auto-electricians report that poorly-made chassis earths are a very common cause of electrical faults. This book advises against the practice. Unless the chassis is being used as a return path there is no need to earth negative leads.

If there are many appliances, take their separate leads from the main power post to a further set of contact breakers, and/or fuses, to sub-circuits. These sub-circuits may (for example) individually feed lights and water pumps. Always feed the fridge via its own heavy cable circuit.

Heavy cable is needed between battery positive and a main circuit breaker, and another between that circuit breaker and a positive power distribution post or connector box. If a current shunt (Figure 4.20) is included, it can usefully be inserted in this lead.

The main negative cable from the battery is taken to a negative power post or terminal strip.

Sizing main battery cables

These cables carry the total current of all connected devices and must be sized accordingly. Even though all appliances may not be in use at the same time, assume this will happen when doing the sums.

If an inverter is included, assess its surge current draw as twice the inverter's continuous rating. If an inverter has a continuous rating of 1200 watts the current drawn is 1200 watts divided by 12 volts: i.e. 100 amps. Assume at least 100 amps and size the cable accordingly.

When sizing cables, allow 50% or so more the start/stop current of the fridge, 6 mm² is not overkill for runs of two to four metres. Likewise for a water pump.

Lighting (particularly by LEDs) usually presents no problems. Do not, however, use less than 1.5 mm² as anything smaller is likely to break, and difficult to ensure that it is terminated firmly within light globe bases and switches.

With some post-2014 or so vehicles, the main cable between the chassis and starter battery doubles as a current shunt, i.e. the voltage drop across it is used as measure of the total current drawn by the vehicle's electrics. Do not interfere with that cable in any way.

Safety precautions

Figure 5.18. The cause of failure with this battery is not known - but an exploded case (and consequent sprayed acid) can happen with any form of lead acid battery- regardless of quality. Pic: original source unknown.

Accidentally shorting battery terminals of a heavy current-carrying cable releases huge amounts of electrical energy. It can vapourise a large spanner or wedding ring, in a (literal) flash.

Vapourising metal is not a clever thing to do as the molten metal inflicts deep and serious burns. In some cases it may blow a battery apart, splattering one's eyes with acid as it does.

Chapter 19

Installing solar modules

Ideally, solar modules in the southern hemisphere should face north, and be tilted toward the sun. Tracking the sun horizontally, but in a fixed vertical plane, may increase their input by 20-30%. At higher latitudes (e.g. Hobart), tracking in both planes may increase it by 40%. The gain in such latitudes (mainly in winter) is thus substantial but, if the array is manually adjusted throughout the day, your risk of falling off a ladder or roof becomes unacceptably high. Automated tracking systems exist, but most are heavy, bulky and costly.

Figure 1.19. Two solar modules are hinged together and have folding angled legs to enable tilting to face directly into the sun.
Pic: Brian Fox (who also made them).

As solar capacity continues to fall in price, excepting for extremely high latitudes, and/or optimising for summer/winter (Chapter 3), the loss in solar input through fixed mounting has long been far more simply (and increasingly cheaply) accomplished by increasing solar capacity.

The solar tables (Tables 1.14 and 2.14 in Chapter 14) show output for both horizontal and tilted modules, but all applications described in this book assume close to or actual horizontal mounting. For much of Australia, horizontal mounting is just fine during much of the year. A minor tilt (for non-mobile installations) reduces condensation and rain water pooling.

Figure 2.19. Self-built folding 120 watt solar modules. Pic: (Middle Lagoon, north of Broome) courtesy of Brian Fox.

Placing modules

Portable modules enable an RV to be in the shade, but are easy to steal. Further, if you have more than one or two modules, the connecting cables required are heavy and unwieldy.

The latter problem can be overcome by connecting the modules in series (for a higher voltage) and installing a suitable MPPT regulator (many accept a range of input voltages) at the RV end. Some have an MPPT regulator inbuilt - but that does not compensate for often too-light cable. Yet another solution is to have some modules permanently mounted and the remainder carried loose.

Yet another solution is to have some modules permanently mounted and the remainder carried loose.

Standard modules are readily hinged via adjustable legs, enabling them to face into the sun at the optimum tilt angle (Figure 2.19). This also enables them to be stored compactly and more safely.

Connecting loose modules

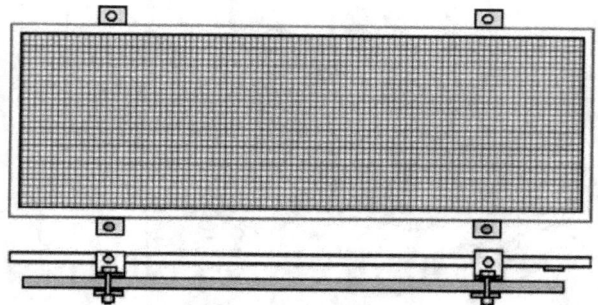

Figure 3.19. Most solar modules have an alloy frame and can be located via aluminium or stainless steel brackets. Sketch: solarbooks.com

The vehicle ends of the cable/s should be terminated by Anderson connectors. These connectors are overkill here, but there's nothing else that will do the job as well at anywhere near the price.

Roof mounting

Roof mounting is cheap, simple and effective. It necessitates the vehicle being in the sun, but having a white, heat-insulated and vented roof keeps the interior within a degree or two of that outside.

Solar modules add further heat insulation, enhanced by the necessary 50 mm or so air space beneath them. Adding a solar-powered fan further increases ventilation. Such fans are stocked by a few specialist suppliers.

Some people have modules roof-mounted but removable when required. These are only for the strong and agile. Typical 120 watt solar modules weigh 10-12 kg and their size makes them unwieldy to move.

Mounting modules on a pop-top roof is likely to necessitate stronger springs or heavier gas struts. Check first with an equivalent weight (e.g. a pillow case full of sand).

Trailers/fifth-wheeler caravans

Here, an excellent solution is to have some of the modules on the roof of the trailer and the remainder on the roof of the towing vehicle. This enables the towing vehicle to remain in the sun and the living part to be mostly or totally in the shade.

Some battery capacity needs to be in the living part so that power is always available if/when the other half is away (see Example 7 in Chapter 15.)

Physical mountings (RVs)

Figure 4.19. Two 80 watt modules run the 60 litre Engel fridge in the author's (since sold) 4.2-litre TD Nissan Patrol. Pic: rvbooks.com.au.

Roof-mounted modules need locating with a 15-25 mm air gap beneath. This keeps them cooler, helps heat insulate the vehicle roof, and prevents water being trapped.

There are various ways of attaching modules to the roof, depending mainly on whether the roof can take screws or bolts, and whether any warranty is invalidated if you drill holes.

Space single modules off the roof using brackets cut from (50 mm) aluminium angle. Multiple solar modules are best attached to a light sub-frame attached to the roof by through-bolts, preferably stainless steel.

Another way, commonly used in the USA, is to attach the sub-frame to the sides of the vehicle. One US RV builder uses tensioned straps secured at chassis level. It is ugly but it works.

Professional installers often attach the solar module sub-frame with marine-grade Sikaflex. This product remains slightly flexible but holds like you wouldn't believe. Removing it requires a chisel and a lot of effort. Before finally Sikaflexing anything together, triple-check the placement, and also that you can subsequently replace or remove the modules.

It is usually necessary to install cabling at the solar module ends before finally securing the modules into position. Ensure you Sikaflex onto something strong - not just a coat of paint!

Flexible stick-on solar modules are available, but they cost more and, if they fail, you are faced with getting them off. They are handy for steeply curved surfaces but it is usually better to stick them onto a thin substrate that can be more readily removed.

If you drill holes in the roof, seal the gaps with high-quality silicon sealant - $3.95 cheapies 'chalk' in the sun. If you stick things down, check that the adhesive is compatible with the receiving surface. Petroleum and silicon based adhesives may affect rubber-based materials.

It may be possible to take cables through an existing vent. If not, an alternative is a cable gland such as Whitworth Marine's Catalogue Number 33516. It is costly but is literally storm-proof. Clipsal connection boxes can be used, but are a visual intrusion on an RV roof. If used, employ plenty of high grade sealant.

Cabins

Solar modules are available as replacements for materials such as tiles and slates. This may be worth considering if you are building and the roof has (or can be given) the required inclination. Having modules at ground level is more flexible and eases installation and cleaning.

Commercial mountings are available, but making your own saves money and is not hard to do. They can be made from galvanised 'C' section roof purlin. The mountings need bolting down to concrete beams about 400 mm by 400 mm, or 500 mm by 500 mm in cyclone-prone areas.

Figure 5.19. Overkill for most locations - this frame, holding six 130 watt modules, is one of seven in the big array that the author built for his previous property outside Broome. It was designed to withstand winds in excess of 240 km/h. The main supports are bolted via hi-tensile studs to massive buried-concrete beams. Pic: solarbooks.com

The array shown in Figure 5.19 is part of the author's own previously owned system scaled to withstand 240 km/h winds on the ocean exposed and cyclone-prone site north of Broome. It has an extra heavy (100 mm by 50 mm) version of 'C-section' purlins and a base (not visible) of reinforced concrete beams, each 600 mm by 600 mm by 3500 mm. The X bracing constrains cyclonic wracking forces.

Interconnection for higher voltage/current

Twelve volt systems are assembled from one or more 12 volt modules. As with batteries, 24 volts can be obtained by connecting two 12 volt solar modules in series.

A single 12 volt solar module that generates 5 amps produces about 60 watts.

Two such modules connected in series still generate only 5 amps, but now at 24 volts: i.e. 120 watts.

Single 24 volt modules are an alternative and are available in higher wattages.

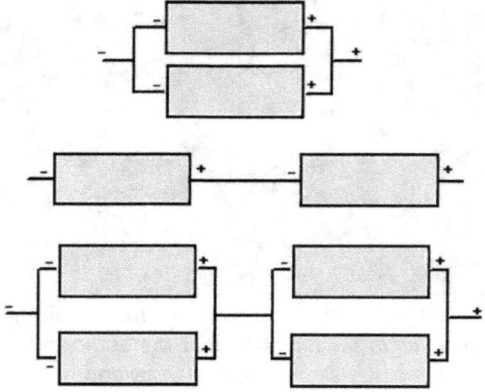

Figure 6.19. Top: parallel connection (current is additive). Centre: series connection (voltage is additive). Bottom: series-parallel connection (both voltage and current are additive). Pic: rvbooks.com.au

You can parallel similar voltage modules of any wattage to increase total current, for example two 12 volt, 5 amp connected modules provides 12 volts at 10 amps - again at 120 watts.

Twenty-four volt systems can have their capacity increased as shown in Figure 6.19).

Forty eight volts is obtainable from four 12 volt modules in series, or two 24 volt modules in series.

There is no 'magic' way to interconnect solar modules (or batteries) to obtain more output. Any combination of similar modules may be connected to provide more amps or more volts, but however done (and like batteries) they always provide the same wattage.

Solar module cable sizing

Overall voltage drop is often 0.3 volt, but 0.15 volt is better and ideally across the whole run from solar modules to battery, and including any cable/s connected via the solar regulator.

The configuration shown in Figure 7.19 is a typical (basic) solar charging system.

The example assumes that cables A, B and C are (each) three metres.

Cable D carries only a fraction of one amp so 1.0 or 1.5 mm auto cable is fine.

Cables E indicate how to parallel connect further modules. The voltage drop of these (E) cables too needs to be taken into account, but only for the current flow to cables A and B.

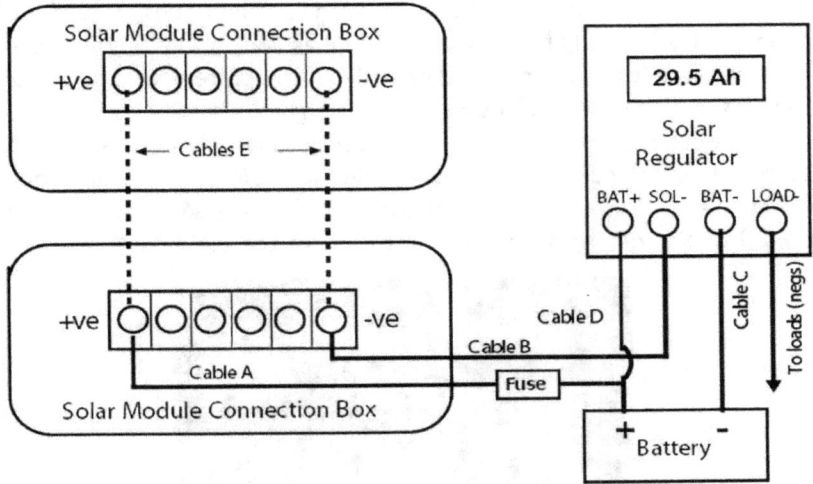

Figure 7.19. Typical solar system connections and recommended (see main text) cable sizes. Pic: rvbooks.com.au

Whilst overkill, use the same cables for (E) as for cables B and D to the lower solar module connection box.

For this example we will assume the total length of cables A and B is six metres and that the maximum current is 10 amps. Using the formula recommended in Chapter 16 and assuming 0.2 volt drop we have 10 x 6 x 0.017 divided by 0.2 = 5.1 mm².

Here, good choices would be 6.0 mm² ISO, 10 AWG, or 10 B&S. Any of them results in a very acceptable voltage drop of less than 0.2 volt drop.

Cables E carry five amps across a metre or so. Whilst it is total overkill it is simpler to use 6.0 mm² cable for this also. Chapter 16 (Table 1.16) has a comparison of the various cable 'standards'.

Series diodes

Some solar modules have diodes (electrical one-way current devices) installed within the junction box on the rear of each solar module, or supplied for buyers to install if they so wish. These diodes (Figure 8.19) are necessary for truly basic systems (such as water pumping) where the solar modules are connected directly to the battery to prevent batteries discharging through the modules at night. Whilst necessary for the above use these diodes introduce appreciable voltage loss, and are also prone to failure. They are not needed in cabin or RV applications as all such systems have a solar regulator that does the same job and with only 0.2 or so volt drop. For such use remove the diodes.

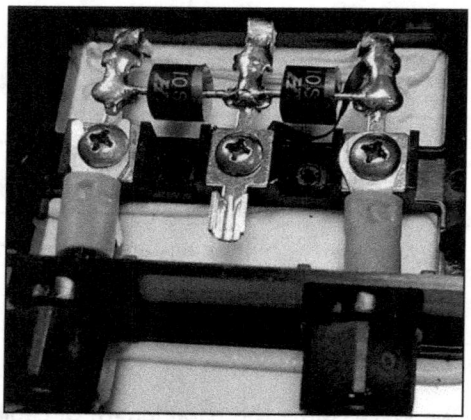

Figure 8.19. Series diodes in a typical solar module junction box.
Pic: original source unknown.

Earthing

Solar modules tend to acquire strong electrostatic charges particularly whilst driving and in windy conditions on hot dry days. Such charges attract dust and in extreme situations may damage the modules electrically.

With large arrays there is also a risk of damage from lightning.

Modern vehicle tyres are partially conductive so the chassis is effectively earthed. With RVs this discharges electrostatic build-up to earth, and lightning damage is reduced or eliminated. Most modules have an earthing point for this purpose.

The frames should be earthed to the vehicle chassis via not thinner than 4 mm² cable. Fixed installation (for cabins) should be earthed via one of those copper (or copper-plated) earthing rods made specifically for this purpose. They are stocked by electrical wholesalers.

Chapter 20

Installing the solar regulator

The solar regulator needs locating where its read-out can easily be seen. A good place is on the outside of a cupboard, with the wiring concealed within. It will need some air flow around it.

Figure 1.20. Plasmatronics PL 20 solar regulator.
Pic: solarbooks.com.au

Regulator wiring is reasonable straightforward, but there is yet another trap that tends to catch out people who know about electrics but not about solar regulators. It is that in order to set the correct charge rate for the battery, a solar regulator relies on knowing the exact voltage across the battery. This is usually done via one or two thin voltage sensing leads.

The Plasmatronic PL solar regulator (Figure 1.20) for example, must have the cable (A) from the solar module positive run directly to the positive battery terminal. A separate lead is then run back from that terminal to the B+ terminal on the solar regulator - as shown in Figure 2.20.

It might seem to make sense (but does not) to run that main solar feed cable *via the solar regulator* to the battery as shown on the 'faulty' connection drawing (Figure 3.20) as this saves using that reference lead. Or, where two such cables are required, as with some regulators, to omit one. Sadly some 'professional' installers do.

If such an error is made, the voltage drop on the related cable between the regulator and battery causes the regulator to 'see' a voltage closer to that of the solar module and thus higher than that of the battery.

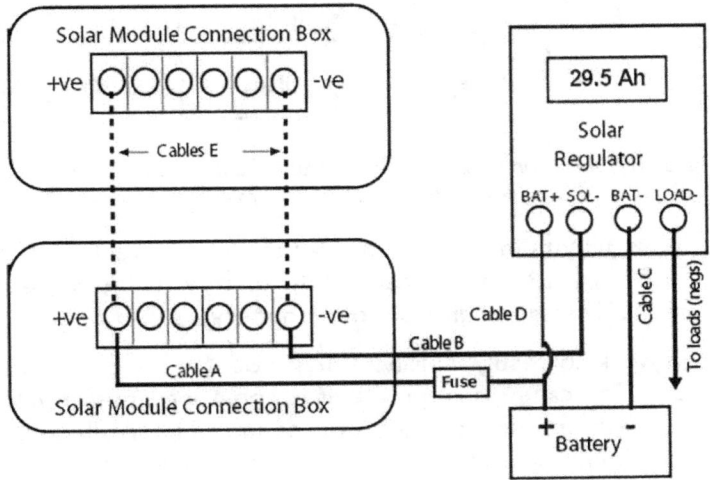

Figure 2.20. Above - correct installation of a Plasmatronic PL regulator. Cable 'A' must be wired as shown (a heavy cable from solar +ve to battery +ve - and a cable from there to the regulator +ve. Pic: solarbooks.com.au

The thinner the cable and the greater the distance from the battery, the greater that voltage difference will be. This higher signal voltage causes the regulator to perceive the batteries as being at a higher voltage than they are and it cuts back the charge accordingly.

Other regulators may require both positive and negative voltage reference leads for this essential function.

If the regulator really is meant to be wired as shown (left) in Figure 3.20, and some cheap ones are, locate the regulator as close as is feasible to the battery, and make sure you have no more than 0.05 volt drop across the cable from the regulator to the battery. Cable sizing for the above (and similar) regulator circuits is covered in Chapter 16.

Load measurement

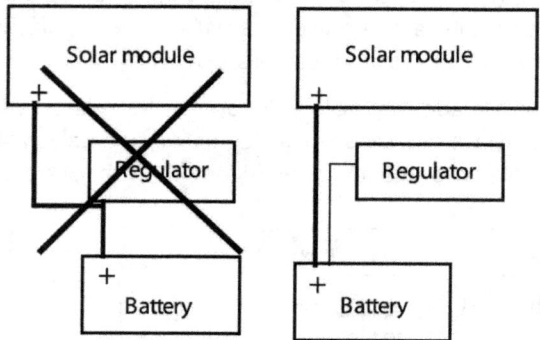

Figure 3.20. Faulty connection of Plasmatronic PL (left), correct connection (right). Pic: solarbooks.com.au

Some solar regulators monitor both solar and load current. This may require returning load negative leads to the battery via the regulator's 'Load' terminal (instead of directly to the battery).

For this to work correctly, appliances must not have any other form of earth return. This can present problems as some appliances, especially fridges with metal frames, have negative connected internally to that frame. If that frame is earthed, their current draw will be recorded as less (or barely at all) because there are now two parallel negative paths - one via the negative cable and the other via the chassis, and only one path is being monitored. This is covered in more detail below.

Most solar regulators only monitor loads within their charge rating. Some are limited to less: e.g. the Plasmatronic PL 20 monitors 20 amps, but the PL 40 monitors only 7 amps and early PL 40s only 5.0 amps. This is a major limitation if one wishes to monitor the current draw of a big inverter, or the input from the alternator, or from a big battery charger. The solution is to use a current shunt to measure that current draw.

Current shunts

Figure 4.20. Typical current shunt. This one can handle up to 200 amps. Pic: rvbooks.com.au

A current shunt consist of one or more short metal strips in series with a main battery lead. A shunt introduces a very slight resistance that causes a small signal voltage to be developed across it. This voltage is directly proportional to the current flowing through the shunt. The signal (typically 50 millivolts at 200 amps) is thus an indirect but accurate measure of current flow.

The shunt is usually connected to the solar regulator (or battery monitor) via an adaptor, typically via a pair of light twisted wires. Ones made recently however may be via a USB cable.

A shunt is not overly difficult to install, but there are a few traps. Unless you are reasonably conversant with electrical practice it is probably better to have an experienced auto-electrician do it for you.

Positive versus negative (shunt) sensing

The shunt may be installed in either the positive or negative battery cable. As mentioned briefly above, if you have it in the negative side of a system that uses the chassis (even accidentally) as a negative return, some part of the current may pass through the shunt, and some through the earth return of equipment that has its negative connected to earth. That current *bypasses* the shunt and is thus not registered. This can occur (much as noted above) with an Engel fridge if it is located by metal tie-downs to the vehicle chassis (its negative is connected internally to its outer metal case).

Where a current shunt is used with an external energy monitor (as opposed to the in-built monitoring facility of a solar regulator), *all* charge and load current must pass through that shunt requiring all inputs and loads normally to be taken to the non-battery side of the shunt.

If however using a shunt in conjunction with a solar regulator that has its own in-built monitoring, you must take the solar feed (only) to the battery side of the shunt. This is usually, but not invariably, made clear in the solar regulator instruction manual. It must bypass that shunt, otherwise the solar input will be registered once when it passes through the regulator and again when it passes through the shunt. The regulator monitor will thus show it incorrectly as doubled. This can result in some users wrongly believing the system has twice the solar input than it actually has.

If desired and where relevant, the alternator charge to the auxiliary battery can also be registered on the energy monitor by taking its battery feed via the shunt.

For systems with variable voltage alternators, all negative return cables (including the auxiliary battery negative) must be connected to the chassis end of the starter battery negative cable - not the starter battery negative terminal. An incorrect connection here results in the auxiliary current draw not being monitored by the vehicle system and may lead to a flat starter battery. See also Chapter 16 and Figure 1.16 regarding better-avoided earth returns.

Chapter 21
Installing the fridge

A fridge's energy consumption, and its ability to cool in extreme heat, depends substantially on how well it is installed. An RV fridge needs installing such that, apart from the door, it is totally sealed from the RV's interior. Not all installers do this properly.

All fridges must be level and out of the direct sun. The wall behind them should be heat insulated. There must be provision for cool air at their base and for this air to be directed over their cooling fins.

Some fridges dissipate heat from their sides. These fridges need a 50 mm air gap either side, plus provision for cool air to flow from their base and up their sides. Whilst not always possible, the rising heated air really needs to be vented to the outside, ideally at roof level.

Details will vary from fridge to fridge. In particular, ensure that cold air is vented such that it flows only through the cooling fins, rather than by-passing them.

Baffles assist a great deal. They can be of cardboard, alloy or timber and installed so they extend to within a centimetre or two of the fins. They are shown as short thin lines in three of the examples below.

Cooling problems whilst running on 12 volts, but less so on 230 volts, are almost always due to inadequate cable size, bad connections, or both. Problems tend otherwise to be caused by poor ventilation, faulty door seals, and (for gas fridges) out of level installation or operation.

Gas/electric fridges must be connected via adequately sized cable and that is much thicker than typically used. Most fridges are have at least 0.5 volt drop and 1.0 volt is not uncommon. Limiting voltage drop to 0.15 V is desirable, with 0.2 V as a maximum.

Achieving this overall voltage drop often necessitates replacing the originally-supplied fridge connecting cable as that *alone* may introduce over 0.2 volt drop.

A further cause of inadequate cabling is that the larger versions of three-way fridges draw a lot more current than many people realise. Even 120 litre units draw 12.5-15 amps. Large three-way fridges may draw over 25 amps (at 12 volts). This necessitates heavy cable.

Example 1

This relates to a compressor-type fridge in a motorhome. The fridge draws 5.0 amps and is sited three metres from the main battery connection. The total conductor length (wiring is twin conductor) is thus six metres. Applying the formula, for 0.15 volt drop gives: 5 x 6 x 0.017 divided by 0.15 = 3.4 (mm^2). Here, 4 mm^2 cable results in a drop of less than 0.15 volt resulting in a very happy fridge.

Example 2

Complications set in where an electric-only fridge is to be run from a caravan battery - yet those batteries are charged from the alternator as well as from solar.

Have as large a proportion as possible derived from solar and as heavy a power cable as feasible between the car and the caravan. Here 13 mm^2 or 16 mm^2 cable is far from overkill.

Better still is to install that heavy cable and also a 12 volt to 12 volt dc-dc converter or charger as close to the caravan batteries as possible. This ensures optimum charging and also provides optimum voltage for the fridge.

Almost all compressor and three-way fridges can have their electrical cooling performance improved and energy consumption reduced by following the advice in this Chapter - sometimes dramatically so.

Chapter 22

Installing an inverter

You may legally do 12 or 24 volt wiring yourself, but if the inverter is to connect into fixed 230 volt wiring, that part of the job must be done by a licensed electrician.

Big inverters may draw as much current as do light starter motors. They need to be housed as close to the battery as is feasible and connected by heavy cable.

To calculate cable size, take the maker's continuous power output (in watts) for the inverter and add at least 50%. For 12 volt systems, dividing by 11 results in amps (and allows for internal inverter losses). For 24 volt systems, divide by 22. Size the cable, so as to result in no more than 0.2 volts drop, using the conversion chart and formula (Chapter 16).

Do these calculations before finalising plans: the cable is likely to be large and you need to make sure there is space for routing it.

Protect the cable by including a manually re-settable circuit breaker or fusible link (Chapter 16) as close as possible to the battery and rated at 30%-40% higher than the peak current draw of the inverter.

Battery/inverter cables must also be protected against breaking, working loose, or being damaged.

If necessary have the cables made up by an auto-electrician.

Before finalising plans, check the inverter's noise level: many have cooling fans. You may need to house the inverter in a sound-insulated (and ventilated) enclosure. Or locate it elsewhere.

Connection sequence

Connect the 12/24 volt side by following the inverter maker's installation sequence. If none is given, follow this sequence carefully:

1. Ensure that the circuit breaker is 'OFF' (or the fusible link is not in place).

2. Ensure the inverter is 'OFF'.

3. Connect the negative (black) lead to inverter negative.

4. Connect the other end of that negative lead to battery negative.

5. Connect the positive (red) lead to inverter positive.

6. Connect the other end of the above lead, via the circuit breaker, or fusible link, to battery positive.

7. Re-check 1-6 and fix immediately if incorrect.

8. If all is well, insert fusible link, and/or click circuit breaker to 'ON'.

9. Switch on and check that all indicator lights etc are as per the maker's instructions. Then plug in a 230 volt appliance and check that it works.

If it does not work, try the following:

Some inverters switch on automatically whenever a load is applied. The threshold of this setting is adjustable and it may have been set too high. Try plugging in a heavier load.

If the inverter now works, reconnect the lightest load you are likely to use and adjust the inverter threshold setting so that the inverter switches on when this load is switched on, and switches off when the load is switched off.

Chapter 23

Installing water systems

Water resists being pumped. The resistance reduces water pressure and flow and wastes a great deal of energy. It is akin to that of current flowing through an electrical conductor, but even more so: e.g. a 12 mm diameter hose needs five times as much power to push water through it as does a 19 mm diameter hose. This is not of huge importance in the average RV but becomes so in large coaches. It can be significant in big cabins and hugely so with irrigation systems.

Water also dislikes turning sharp corners. Where feasible have sweeping curves rather than right-angle bends. Use food-quality hose for drinking water and water for cooking, secured via stainless-steel hose clips.

Water pumps draw about twice their running current (doubling their typical four to six amps) for a second or two whilst starting. The supply cable should be rated for this starting load (i.e. eight to twelve amps), as pumps are prone to stall and overheat if voltage is low: they may even burn out. Install a blade-type fuse close to the pump. Try a 10 amp slow-blow type first. If this blows frequently, replace it by a 15 amp normal fuse.

Most RV pumps have a pressure switch and are connected as their makers show. A few pumps (made mostly for irrigation) need an external pressure switch. These switches can be located anywhere between the pump and the tap/s.

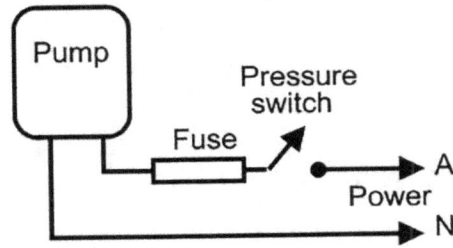

Figure 1.23. Installing a remote pressure switch.

In either case include a remote on/off switch in a readily accessible location so that the pump can be turned off if necessary for any length of time. Or feed it via a circuit breaker that doubles as that remote switch.

Replacing an internal pressure switch costs much the same as buying a new pump. If that pressure switch fails it can be bypassed and replaced (externally) by a far-better one made for irrigation use. They last forever and cost only a third or so the price of a replacement pump. They can be adjusted to work over the required range: 140-350 kPa (20-50 psi). Connect as shown in Figure 1.23.

Pressure tanks

Some pressure tanks have a separate water inlet and outlet, necessitating two corresponding hoses. Others have a single outlet and require a tee connector.

With either, the pressure tank, pressure switch and (optional) gauge can be located and teed off anywhere that is convenient between the pump and the first tap. It is just fine, for example, to have the pressure tank and pump underneath and at the rear of the vehicle (or outside a cabin), and the pressure gauge in the kitchen connected via a small diameter pipe (any size over a few mm will do).

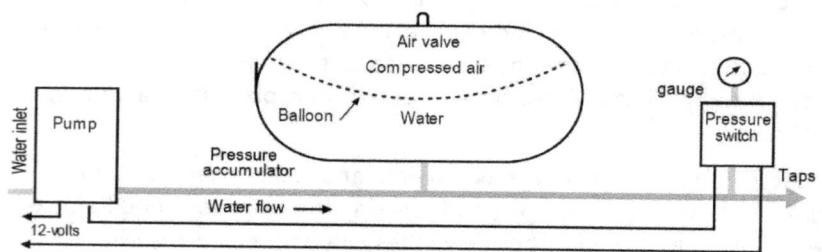

Figure 2.23. Pressure tank installation is very flexible. The tank, pressure switch and gauge may be teed off anywhere between the pump and the first tap.
Pic: rvbooks.com.au

Chapter 24

Installing a voltage sensing relay

Where a fridge or the RV system is to be powered by a battery charged and by a conventional alternator (i.e. any that never drop below 12.7 volts whilst driving), it is essential to set this up so that the starter battery is assured of starting priority, and is protected against accidental discharge.

Figure 1.24. Voltage switching relay, links batteries only when the starter battery is adequately charged. Pic. redarc.com.au

Until the 1980s or so this used to be done manually by a heavy current switch, or via a basic relay that linked batteries when the ignition was 'on' and separated them again when it was turned off. Neither, however, provided automatic starter battery charging priority. This could (and often did) result in the starter battery being discharged by a 'flat' auxiliary battery suddenly being connected - or by forgetting to turn it off.

From about 1980 onward the above issue was overcome by a voltage sensing relay that isolated the auxiliary battery from the starter battery (and alternator) until the starter battery recharged, typically inside a minute or two. The relay separated the starter battery again if its voltage dropped below the 12.5-12.7 volts needed for cranking.

The example shown (Figure 1.24) has an optional connection that enables the auxiliary battery to be paralleled across the starter battery manually to aid starting (if extra current flow is required).

Another recently introduced unit, from Intervolt, is programmable for different pull-in and dropout across a useful range of voltages.

Variable voltage alternator issues

These relays cannot be used with the mainly post-2013 vehicles that have variable voltage alternators as their voltage drops below the relay's hold-in voltage every time the brakes are applied or the accelerator pedal released whilst (say) descending a long hill (Chapter 6).

Vehicles with these alternators can only charge an auxiliary battery by using a specialised battery to battery dc alternator charger. These have starter battery charging protection inbuilt.

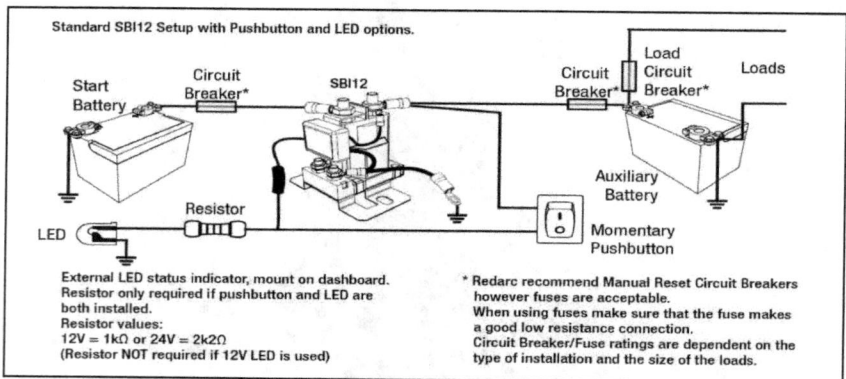

Figure 2.24. How a typical voltage sensing relay is connecte. The (blue) cables, push button, resistor and LED are needed only if an auxiliary battery is to assist starter battery. Pic: redarc.com.au.

Chapter 25

Electrical converters

Most RV makers worldwide assume that owners of their products will stay in caravan parks (where grid power is available) with only an occasional night stay away from grid power. The RV's electrical system is designed accordingly. Whilst fine for its intended purpose that system presents problems if grid power is not available for more than one night. Modifying it, however, can be a major job.

The systems are all based on variants of a so-called 'electrical converter'. This is a large and usually heavy step-down transformer and rectifier that runs from 110 or 230 volt grid power and supplies the RV's 12 volt system with 30-50 amps at an unregulated (nominally) 13.65 or so volts dc.

All of an RV's 12 volt bits normally run directly from this converter via that grid supply. A typically small 12 volt battery is included, but that battery does not supply power in normal usage. That battery is automatically switched into use only in the absence of grid power. (In the USA, that battery is often designated 'Emergency Power').

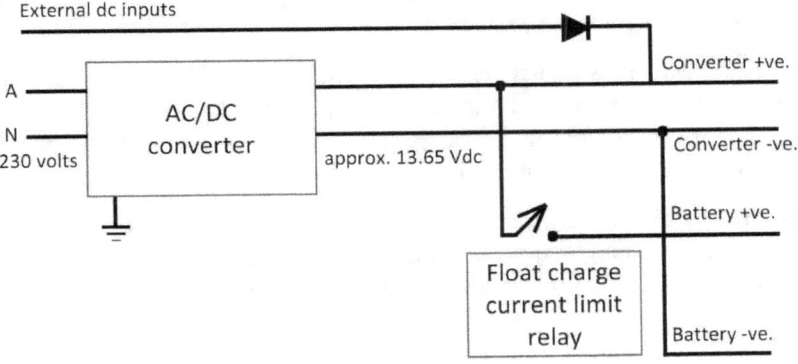

Figure 1.25. Typical basic RV converter. Pic: rvbooks.com.au

Few converters have a conventional battery charger as such. They use the converter's 13.65 volt output to maintain charge in a battery that is presumed already charged and also protect the battery between infrequent RV uses (as routine in RV hire fleets).

That 13.65 volts, however, is far too low for meaningfully charging even the small battery typically supplied. It will partially charge the original battery but will take a day or more to do so, but cannot cope with the 150 to 300 amp hour capacity required for free-camping.

It is feasible to replace the converter by a good quality three-stage battery charger run directly from the grid supply. The battery too will need to be replaced, but this often presents the problem of where to find space to locate the much bigger one/s required. Structural reinforcement may be needed to carry the considerable extra weight - unless a (much lighter) lithium battery is used.

There is, however, likely to be a further associated problem.

Converter associated wiring issues

A converter's output of about 13.6 volts is almost one volt higher than from a 90% charged deep-cycle battery. Many RV makers take advantage of that higher voltage to use thinner cable, on the basis that even a 1.0 volt drop then only marginally matters. This does not matter whilst grid power is available as appliances still have an adequate 12.6 or so volts.

That which does matter, however, is that replacing that converter by a high-quality battery charger will then fully charge the battery, but cannot compensate for that cable voltage drop whilst running from battery power. Appliances will still have totally inadequate voltage (such that a compressor fridge will barely work).

A partial solution is to use lithium (LiFePO4) batteries. These charge to 80%-85% from 13.65 volts and rarely drop below 13 volts in typical RV use - but it is still a compromise as this still results in appliances running on a lower voltage than intended.

The only total solution is to upgrade all main charging circuit wiring, plus wiring to all high current loads. LED lighting wiring, however, is usually fine as LEDS are not that voltage sensitive.

Chapter 26

Compliance issues

As advised in earlier editions of this book, vehicle regulatory authorities (especially in Queensland) have tightened up their inspections, particularly of self-built RVs and imported fifth-wheel caravans and motorhomes to check whether they are fully compliant with Australian requirements.

Apart from the possible electrical issues (that an RV may not have a certificate of electrical compliance, until 2011 or so, some inspecting authorities accepted written declarations that the required Standards had been met, on the reasonable assumption they were valid, however many proved not to be. This resulted in their being approved as complying and registered accordingly.

Worse, and as later reported in Federal Parliament, many imported prior to 2010 were found to have false or forged certification.

Further, as detailed below, a substantial number of those privately imported are only partially compliant, and whist that enables the original buyer to use them in that form, they cannot legally be sold, let alone registered, until made 100% compliant.

Readers planning to purchasing any self-built RV, and new and (particularly) second-hand imported fifth-wheel caravans are very strongly advised to make 100% sure of this as many of their owners truly believe they are fully compliant - but few are. A complete giveaway is if any 110 volt anything is installed or included. If they have, the unit is not compliant.

This raises major problems when such vehicles are presented for re-registration. It is now totally known by regulatory authorities that some compliance certificates are invalid. Most now check these units very thoroughly, particularly their electrical and gas installations and also that their overall width and rear overhang do not exceed legal limits.

Although assurance of compliance may be written into the sales agreement, even that may not guarantee that all required work has been or will be done. Discuss the electrical requirements with a local licensed electrician experienced in RV work (the requirements differ substantially from domestic practice) and follow the advice. Then have that electrician physically check the vehicle. Have a similarly experienced gas fitter check that side of the work.

Have the vehicle weighed in your presence on a certified weighbridge and obtain a printed weight certificate. Do not accept any previous certificate, no matter where from.

The major concerns are non-compliance with the stringent requirements of AS/NZS 3000:2018 and AS/NZS 3001:2008 relating to mains-voltage wiring and related obligatory protection devices; gas installation, overall width (including any awning), rear overhang and often gross overweight.

The reason to take this so seriously is that the necessary work to ensure compliance can be hugely costly. It can mean total rewiring, replacing springs and shock absorbers and even the entire braking system. If over-width it may be impossible to achieve compliance and be thus rendered not legally salable.

Whilst the vendor that sold the unit to you is legally liable, at least one went into receivership to avoid payment.

Whilst outside the scope of this book, there are also likely to be issues of gross overweight and suspension that is unsuitable for Australian roads.

Electrical non-compliance

Most major non-compliance issues are now with RVs that have been privately imported. Even if buying via a 'facilitator' the legal buyer is the person who pays the bill.

By grossly exploiting a legal loop-hole that enables overseas visitors personally to use 110/120 volt razors etc in Australia (but not import them for sale), privately imported RVs are 'converted' by adding a 230 volt to 110/120 volt transformer, so the RV's existing wiring and appliances can be used. This allows the original buyer (only) to legally use that vehicle but does not however confer electrical (or other) *compliance*. Such RVs cannot legally be sold (nor even given away) unless brought into 100% compliance.

Very few owners of these vehicles are aware of this. Most believe totally, but wrongly, that their RV is 100% compliant. It rarely is. The reason why it cannot be compliant is because, in Australia and New Zealand, all RVs must meet relevant Standards. None allow any domestic 110/120 volt appliances to be sold in either country. Such appliances are in any case designed for 60 Hz operation (not 50 Hz) and may overheat or misbehave in any manner of ways. Nor can the existing wiring, switches or power outlets be used to carry 230 volts.

In such cases the RV's entire 110/120 volt system (including cabling circuit breakers, switches, socket outlets etc.) must be replaced. This may require the removal and subsequent replacement of the RV's entire outer skin. The invariably installed 110/120 volt ac to 12 volt dc converter must also to be replaced by a 230 volt ac to 12 volt dc converter.

Chapter 27

Fixing problems (general)

Apart from the converter issues described in Chapter 25, a probable 90% of all other 12/24 volt electrical problems are directly due to too-thin cable, loose or corroded connections and buying almost useless ultra-cheap products. Avoid or fix these problems before adding solar or you will not know where lies the cause of any future problems. Other common issues are faulty fuse holders (see below re fuses) and poorly-made earth connections in systems that have earth returns via the vehicle chassis.

Fuses blow after a few hours, yet there is no apparent fault

The fuse rating should be 50% higher than the maximum current normally drawn, or 75% higher where it is very hot (as in central Australia or above the exhaust pipe, and especially if both).

Electric motors draw high initial currents and need 'slow blow' fuses.

The cause may also be loose or corroded fuse holders, or (and often) poorly crimped fuse holder lugs heating up the fuse and thus causing it to blow. Check for heat under charge or load. Ideally replace all fuses larger than 15 amp, and also their holders, by the high-quality (and much physically) larger units shown in Figures 6.16 and 7.16 in Chapter 16. There is also a known problem with low quality blade fuse holders partially melting or burning out.

Batteries only last a year or so

This is usually due to chronic overloading, but can also be a result of voltage drop as a direct result of the faults described in the item above, resulting in chronic under-charging. Overcharging can also wreck batteries (especially AGMs and LiFePO4s) but is otherwise uncommon except where they have been left permanently connected across a cheap mains charger.

A battery's life is also a function of its temperature. Where possible, locating an auxiliary battery under the bonnet should be avoided. Likewise it is essential to keep batteries out of the direct sun.

Batteries appear to fully charge but are flat soon after a light load

This is a now largely historical problem with earlier battery technology. It almost always indicated that the batteries were badly sulphated, i.e. inactive material flaked off the plates and dropped to the bottom of the cell. This slowly built up and eventually short-circuited the plates. If/when it did, the result was almost instant battery failure.

Such sulphation was caused by age, chronic under-charging, or leaving batteries discharged over time. It showed up as a high charging voltage after some minutes, or sometimes longer, even though the batteries remained deeply discharged. There was an apparent charge, but it existed mostly on what little remained of the surface of the plates. The battery voltage may well have seemed fine for a few minutes but then dropped rapidly.

If you buy only good quality batteries from known major battery makers you are unlikely to have this problem.

Solar charge reduces in the afternoon

This is normal. As batteries increase in charge, the solar regulator reduces the charging current accordingly. With most well-designed systems this reduction typically happens around 2-3 pm on sunny days.

Solar charge is sometimes well above the modules' claimed maximum output

This is common with installations close to the sea, along inland lakes and sandy areas etc, during times of bright sun and also low scattered sun or haze. In such areas, the solar modules receive direct sun via gaps in the cloud, plus sunlight reflected upward from the water or sand and then down again from the underside of white cloud or haze. This typically lasts for a few minutes, but, during that time, the solar input may increase by 20-30%.

Solar charging appears be twice that expected - far higher than that claimed by the solar modules' vendors. Do I have an exceptionally good installation?

No! The solar energy monitor has been incorrectly connected and is being recorded twice. This is a rare but known issue that usually arises where the monitor is part of the solar regulator and registers and display the solar input as it travels through that regulator. If the system also includes a current shunt and the solar input is routed (as with other inputs and outputs) through that shunt, that solar input is registered there also and added to the existing readout. In your installation, the solar input must be connected directly to the battery, not via the shunt. See also Chapter 20 regarding this.

LED globes fail prematurely

LEDs normally have many years life. The usual cause of apparent failure is tarnished pins. Cleaning with fine emery paper usually fixes them.

Overcharging batteries from solar modules with regulator connected

Twelve volt solar modules may generate over 20 volts, but the regulator reduces and adjusts this for optimum charging. Given an adequate and correctly adjusted solar regulator, a system may be left running for years on end. Some makers prefer stored AGMs to be left fully charged, and then charged again only when they have fallen to 60% or so, typically after 6-12 months. LiFePO4 batteries are best left 50% charged - and can then be left like that for many years.

I live in a dusty area yet cleaning and polishing the solar modules' glass exterior covering causes the dust to build up quicker and even worse

You are adding to the problem! Polishing the glass creates a static charge - as does dry wind blowing over the glass. Check that the metal frames of the modules are all thoroughly earthed as that will help dissipate the charge. When cleaning the modules, wash them using water containing a teaspoon or two of detergent (an anti-static agent); then rinse them with clean water that also has a teaspoon of washing up detergent. Allow them to air dry. Never polish them.

I am bothered about a TV program in which a participant claimed that live solar arrays can kill people. How can I protect against this?

That program segment related to risk with grid-connect systems where solar arrays may run at several hundred volts. But most cabin and RV solar arrays run at only 18-30 or so volts dc. Such low dc voltages may give a slight tingle but the main risk is of falling off the roof in surprise.

Chapter 28

Living with solar

Now that virtually all rechargeable batteries likely to be used in cabins and RVs are sealed, a well designed and installed such solar energy system needs little attention.

Solar modules need not be squeaky clean. Even when mildly dusty they lose only a few percent or so of their output. In non-polluted areas, occasional rain does the job, otherwise (unless they are very dirty) use a bucket of clean warm water and a few drops of detergent. Rinse with clean water - and leave to air dry. Do not wipe them dry, and particularly do not polish them. Doing either builds up a static charge that attracts and retains dust. (See also the related item in Chapter 27.)

A correctly working self-sufficient system may not work as you might at first expect. It may not, for example, appear to produce any more power on long bright summer days than in spring or autumn.

What is happening is that during the summer, more energy is available than you may need. If it is not used, the batteries will become fully charged earlier in the day. When this happens, the solar regulator reduces the charge.

Typical indications of a good system with lead acid or AGM batteries

Before sunlight strikes the modules, and with nothing switched on, the battery voltage of a self-sufficient solar system working correctly is likely to be 12.5-12.6 volts. As the day progresses, the battery voltage will rise. Once the sun is fairly high in the sky, the charge rate will remain substantially constant. When the maximum preset charging voltage is reached (which may be as high as 15 volts), the regulator will either hold it at that for an hour or two, or more likely reduce it to about 14.4 volts.

If the latter voltage, the charge current will fall to somewhere between half and two-thirds of that charge current previously. This period (called absorption) typically lasts for a couple of hours, by which time the batteries will be very close to 100% charged.

Figure 1.28. Part mobile home, part mobile office, this converted coach (owned by Eric Brandstater) has one of the best self-installed solar systems I have yet seen. It puts most professional jobs to shame. The coach travels Western Australia most of the year-around. If you see it (it is bright yellow) ask to have a look. Pic: Author 2006. (One week after this photo was taken, in 2006, a bush fire wiped out all the background for about 30 km). Pic: rvbooks.com.au.

The charging voltage then reduces again, this time to 13.2-13.6 volts depending on temperature and battery type. This is known as 'floating'. Ideally this charge balances the battery's internal leakage and compensates for minor loads such as electric clocks. Sealed lead acid deep-cycle batteries can be left on float-charge indefinitely via a high-quality solar regulator.

The lowest voltage indication (usually during the evening) is likely to be around 12.0 volts whilst the batteries are under normal loads. A microwave oven will however (and for some length of time) pull battery voltage as low as 11.4 volts. This is fine as long as the voltage comes back to over 12 volts within an hour or so. A typical deep-cycle battery is 50% discharged at 12.25 or so volts - but that will be its voltage after resting for 12-24 hours. It is likely to be lower under even a light load. Be aware that any instantaneous voltage reading of a deep-cycle battery not only has little meaning, it may seriously mislead.

As a very rough guide, charging is usually satisfactory if the on-charge battery voltage reaches 14.4 volts in typical RV usage by early to mid-afternoon on most days. The system should also go into 'float mode' shortly after. If it doesn't, check that the regulator is set up correctly. If it is, and the batteries are known to be in reasonable condition, increase solar module capacity.

Typical indications of a system with lithium batteries

Lithium batteries have an obligatory management system that ensure they are always correctly charged. As Chapter 5 explains they do so in a manner that it is totally different from other batteries - in that there is only a small voltage difference between 90% to 20% charged.

Their state of charge that reflects the solar input and working of the system can only practicably be gained via an energy monitor that measures current in and current out.

Solar capacity versus battery capacity

As stressed also elsewhere in this and my other books and articles, many older cabin and RV systems have too little solar capacity and too much battery capacity because solar capacity was far more expensive.

Ideally add more solar, but if that is not feasible and battery capacity is truly excessive, it is often beneficial to reduce it. The reason is that large lead acid battery banks have substantial internal losses, so some part of the otherwise stored energy is lost. A smaller battery bank has a better chance of being fully charged.

Training visitors

Early editions of this book emphasised the need to train city-bred visitors to turn off unnecessary lights, take short showers, leave their hairdryer in the travel bag, turn the TV/DVD (or other such electrical acronyms) off at the power outlet, and forgo using the microwave to thaw out frozen chickens - otherwise power usage is doubled, or more.

Experience over the years however shows that asking that energy be conserved does not work with almost all city-bred teenagers, many city-bred people and almost all caretakers. At best, a 15 minute shower becomes 'only' ten minutes.

We attempted to resolve it by setting up our visitor's/caretaker's cottage with its own solar system that automatically shut down if sustained current draw exceeded a pre-settable level, or battery charge fell below 55% remaining. As this happened with most visitors we finally gave up - and installed a quiet generators.

We now live in Sydney, but retain our previous mode of electricity usage - resulting in our house running on a now 6.5 kW solar (and Tesla battery) grid-connect system (and solar water heating). The solar system produces a virtual 100% or our draw in winter, and we feed substantial into the grid for the rest of year.

Now, when visitors stay, the power grid that gets thumped in winter, but we still have solar to spare in summer.

Chapter 29

Walking the walk

Figure 1.29. The author and the QLR in El Golea oasis (Sahara) in 1959. Pic: Tony Fleming.

My interest in much of this goes back to the mid-1950s whilst working in the research division of General Motors. I had been attempting to simulate rough road conditions - but not enough was known about road conditions encountered in Africa (the main export market). With the backing of Mobil Oil, I elected to find out myself.

I used a rare QLR Bedford 4X4 truck that had been built in WW2 as a mobile airfield control centre, but had never been previously used. It had a huge (350 amp) transfer-box driven dynamo - ideal to drive the electrics required. Fuel storage was increased to 1270 litres - a necessary range of 3500-4000 km.

A colleague (Tony Fleming) and I drove it twice the length and breadth of Africa, including the two Saharan crossings. The trip later became semi-famous. Africa was falling apart politically at the time. The Sahara route was closed the very day we completed the return crossing (in 1961) and has never reopened. That QLR was thus the very last vehicle across that central Africa route, all subsequent such drives have to be via the Sudan. A superb 43 minute video of this journey can be viewed at https://www.youtube.com/watch?v=PtG6niRiRXk.

VW Kombi

Figure 2.29. Our Kombi in camp at nightfall, about 200 km north of Birdsville, Qld, in 1996. Pic: rvbooks.com.au

In 1994 my wife and I rebuilt a 1974 Kombi adding a 100 Ah auxiliary battery charged by the alternator, plus a 100 watt solar module. This ran a 32 litre Engel fridge and minor lighting. It worked reliably but lack of monitoring back then made it hard to gain data or track down failings.

OKA

Figure 3.29. OKA crossing the Wenlock River (Cape York). The roof-dome houses the antenna of the Westinghouse satellite phone. Pic: rvbooks.com.au

My wife and I later rebuilt a 1994 OKA (an Australian-built ex mining vehicle) as a go-anywhere mobile home. We drove it many times across Australia, over mostly inland tracks and assisted by a 2500 km safe fuel range.

Aided by two 80 watt solar modules, a 140 amp Bosch alternator initially charged three 100 amp hour deep-cycle batteries.

This drove a 71 litre Autofridge, a laptop computer and printer, ten halogen light globes, and a huge (suit-case sized) Westinghouse satellite phone (the dome antenna of which can be seen in Figure 3.29.). Later, an Iridium sat-phone enabled running on one battery and solar alone.

As with owners of many OKAs we had a number of initial problems with the original (Lucas) alternator, and the suspension. Once sorted out, however, the OKA proved extraordinarily reliable.

Nissan Patrol and TVan

Our subsequent 2005 4.2 litre TD Nissan Patrol had a permanently 'on' 60 litre fridge, driven by two 80 watt roof-mounted modules via a PL40 Plasmatronic solar regulator (and later, and on long-term trial, the very first Redarc BMS 1215) and an 105 amp hour AGM battery.

Figure 4.29. Nissan Patrol and TVan each had its own solar modules, regulator and battery. Its provision for alternator charging was never needed. Pic: Author - Mitchell Falls 2009.

The fully off-road TVan camping trailer had one 50 watt roof-mounted module that charged a 100 Ah trailer-located AGM battery via its own solar regulator (it had no alternator feed).

This basic system powered three LED lights, a water pump and a NextG Blue modem and a laptop computer. It proved to be totally reliable. This concept of twin but separate solar systems works superbly.

Home and property

Figure 5.29. Our own designed and built solar system generated up to 17 kWh a day. Pic: solarbooks.com

In 1999 we bought 10 acres of virgin bush adjoining the Indian ocean north of Broome. Our house and property ran from solar alone. The self-built house was even constructed using almost only solar power.

The solar array had thirty 120 watt and 130 watt modules running at a nominal 72 volts into an Outback Power MPPT 60 amp regulator. They charged 16 gel cells, (960 Ah) at 48 volts (14-17 kWh/day.)

An SEA 48 volt inverter ran a 500 litre fridge, dishwasher, washing machine, two computers, 104 cm TV, kitchen appliances and thirty compact fluorescent globes.

Swimming pool

Re-circulating the water in the 31,000 litre swimming pool would normally require 3-5 kW/day and a $50,000 plus system if using normal (230 volt ac) technology.

*Figure 6.29. The pump's solar array just visible bottom right.
Pic: solarbooks.com*

We installed a 48 volt dc brushless Lorenz motor pump driven directly (i.e. no batteries) from four 120 watt modules via a Lorenz 48 volt MPPT regulator. It re-circulated 3000-5000 litres an hour between sun-up and sun-down. Including the cyclone-resistant solar mounting, it cost just under $7500.

Water pumping

Figure 7.29. Our previously-owned Broome property. All structures were self-built. The land fronts a hairpin-shaped tidal lagoon. The Indian Ocean is at the far side of the sandy bank and beach. Pic: solarbooks.com

All washing and toilet flushing water was pumped from a 100,000 litre rain-water tank that comfortably filled about half-way through the yearly wet season.

Water was pumped to the house by a 750 watt Grundfoss centrifugal pump that initially ran each time water was drawn. It used about 1 kWh/day, but adding a 450 litre pressure tank reduced that by an extraordinary 95%. It subsequently ran only once or twice a day - for about three minutes each time. Water was thus delivered almost entirely smoothly and silently by the tank's air pressure alone. We sold this property in mid 2010 and moved back to Sydney to be close to our family.

Church Point

Home is now an environmentally designed house in Church Point (NSW) overlooking Pittwater. It is north-facing and partially recessed into a steep hillside. Most of the rear is underground. All walls are Hebel (aerated concrete block) or reinforced concrete. The resultant thermal mass keeps the house between 17o to 19°C throughout winter. It needs only minor heating.

We initially installed a 2.4 kW grid-connect solar system that fed 40% of its output into the grid during winter. Realising that, we installed three Daiken reverse cycle air conditioners that, in winter, run on their ultra-efficient heating cycle all day to heat up the thermal mass. Each provides over 3.5 kW of heat for under 700 watts drawn but are rarely needed at that level. Summer temperature remains at a constant 23°C to 24°C excepting for the ground floor vestibule which stays at around 20°C.

In 2019 we increased solar capacity to 6.5 kW and added a 14 kWh Tesla battery. We evaluated going off-grid and converting the system to stand-alone operation with substantial battery backup but felt preferable to retain the grid supply as a 'virtual battery' to cope with rare periods of inadequate sun. During mid-winter we generate marginally more than we use, but in summer we sell 20-30 kWh/day to the grid supplier. When electric vehicles that have at least 750 km range become available, we may use that to provide the necessary source of charging for our limited driving.

Figure 8.29. Our current home in Church Point (north of Sydney).

Chapter 30

Electricity - simply explained

What is electricity?

Lord Kelvin - source unknown.

In the mid-1850s, the great scientific pioneer, Lord Kelvin was lecturing on electricity. He asked his class: "What is electricity"?

One student put his hand up, but then stammered out that he'd forgotten. Lord Kelvin turned slowly to the class, and said:

"Gentlemen, you have just witnessed the greatest tragedy of this century. Only two people know what electricity is. One is God, and the other one is Mr Smith. God won't tell us - and Mr Smith has forgotten".

To this day no-one truly knows, but whilst its exact nature is yet to be fully defined, its behaviour has been surprisingly well understood since the days of Lord Kelvin and many other scientists of his era.

General

Alternating and Direct Current: the movement of electrons responsible for the flow of electricity. Direct current flows and performs work in a manner analogous to a band saw: it operates in a continuous direction. Alternating current works much as big cross-cut saws that operate by being pulled to and fro. Preferred abbreviations are now ac and dc respectively (they were previously AC/DC).

Electricity authorities supply alternating current (although Broome and a few other Australian country towns ran on dc until the 1960s). Batteries supply direct current. It is readily possible to convert alternating current into direct current - and vice versa. An inverter converts direct current into a higher voltage alternating current, a mains battery charger converts alternating current into a lower voltage direct current.

Solar panels/modules: The solar industry describes individual solar generating units as solar modules, 'solar panels' are assemblies of modules. Assemblies of modules are called 'arrays'.

Standards: In all Australian/New Zealand Standards 'should' is a suggestion or recommendation only. 'Shall' is a requirement that must be followed to comply with that Standard.

Electrical units and terms

Amps: the amount of electrical current that is flowing. It is akin to flow in a pipe. The greater the voltage, the greater the amount of current that will consequently flow. Its common abbreviation is A, (but that of current generally, when used in a formal equation, is I).

Amp hour: the amount of electrical current that flows in one hour. A device that generates four amps for five hours thus produces 20 amp hours: amp hour is commonly abbreviated to Ah.

Extra-low voltage: voltage not exceeding 50 Vac, or ripple-free voltage not exceeding 120 Vdc.

Low voltage: voltage exceeding Extra-low voltage, but not exceeding 1000 volt ac, or 1500 volt dc.

High voltage: any voltage (ac or dc) that exceeds Low voltage.

Ohms: The unit of resistance to the flow of an electric current. The unit is either spelled out (i.e. as ohm), sometimes expressed as R, and (traditionally) also expressed by using the Greek symbol for omega (Ω). One ohm is the equal to the resistance to the flow of current through which a current of one amp will flow if one volt is applied across it.

Ohm's Law: a fundamental electrical law. Volts, amps and ohms are interrelated and defined by Ohm's Law - that states that direct current (dc) that flows in a circuit is directly proportional to the voltage across that circuit. It is valid for metal circuits and those liquids that are electrically conductive.

Power/Energy: These terms are often misused. Power relates to the rate at which work is done and is expressed in watts. Energy relates to the amount of work done and is expressed in watt hours.

Resistance: To varying extents, all substances resist the flow of electricity. This resistance generates heat. Resistance can be useful (it is how an electric kettle works) but where heat is not specifically required, it wastes energy. The thicker a cable, the lower its resistance - and the less the energy lost through heat. The term 'resistance' is itself often abbreviated as R. It is measured in ohms.

Volts: the pressure that causes electricity to flow: akin to pressure in a pipe (abbreviated V). It is common to indicate whether such voltage is ac or dc - e.g, Vac or Vdc. In formal equations it is E.

Watts: Volts, amps and ohms are interrelated and, when multiplied together are a measure of energy used, and also of work performed. The resultant unit is a watt (abbreviated as W). Thus one volt times one amp equals one watt.

Watt hour: a measure of electricity generated or used in one hour. A 100 watt globe that is running for 30 minutes consumes 50 watt hours. A 12 volt solar module producing four amps for five hours produces 4 (amps) x 12 (volts) x 5 (hours) = 240 watt hours. The correct abbreviation is Wh.

Watt hours/day: the number of watt hours consumed in a 24 hour period. This unit is handy when scaling solar systems. The correct abbreviation is Wh/day.

Chapter 31
Making contact

Early editions of this book included supplier contact details that rapidly became out-of-date. The ease of locating via Google and the Internet removed this need. Now, many companies prefer only to use email. There is also a proliferation of local and overseas on-line vendors, and many solicit only internet sales.

'Googling' what you seek will return many links, some of which may not be current or exactly what you are looking for, but will certainly be superior to a static list. For these and other reasons this book no longer lists contact details.

Companies can often be located via their trade name followed by .com.au (for some Australian companies). Others have only .com - or .net. There are also organisational suffixes such as .org.

(Where the company name is more than one word try joining the words together, omitting upper case letters, and spelling out '&' as 'and'.

Where the above does not work try using a hyphen: e.g: eastpenn-deka.com. Another way is to search using the on-line Yellow Pages.

Useful sites

Our website (rvbooks.com.au) contains many articles over a wide range of topics. All were totally revised and updated in late-2019 and are now typically updated monthly. It also has Links to other sites of interest. An associated website (solarbooks.com.au) was launched in September 2018.

An excellent example of a really well-run source of specialised information is the camper trailer site campertrailers.org. For 4WD owners, caravans and related topics, I recommend exploroz.com.

Some useful information can be found on web forums, but much is worthless and sometimes downright dangerous. It is mostly of value if you know already know about the topic.

Our other published books (all now in eBook form) are:

Solar Success. This book, a companion to Solar That Really Works, covers every aspect or installing and using solar in home and property systems. It is valid globally.

Caravan and Motorhome Electrics covers every aspect of that topic. Whilst written in plain English and intended for non-technical readers it is also used as a main text by educational colleges in Australia, and many auto-electricians. (A print edition of this book is still available).

The Caravan and Motorhome Book primarily covers Australian RVs but has many overseas readers.

The Camper Trailer Book likewise primarily covers those used in Australia (a major market).

Why Caravans Rollover - and how to prevent it is now our best seller. It is unique in explaining (in plain English) exactly how and why this happens -and (vitally) what owners can do to prevent that happening. It contains an invaluable guide to self-assessing one's own tow vehicle and caravan. It also has a fully technical final part (primarily for engineers).

All our books (and website articles) are routinely updated each year (and sooner if required).

This book's publisher

Following three years in the RAF as a ground radar engineer and two years at de Havilland Propellers, Collyn Rivers joined the newly-opened Vauxhall/Bedford Chaul End Research Centre in the UK.

In the 1960s he drove a big 4WD Bedford QLR twice across Africa, studying track surface conditions.

Collyn then moved to Australia, initially designing everything from X-ray scanners to 500-tonne concrete presses, and then three years as Applications Engineering Manager of Natronics Pty Ltd.

In 1970, he founded the worldwide magazine Electronics Today International (ETI), which, in 1976 was awarded the title of 'Best Electronics Publication in the World' by the Union International de la Presse Radiotechnique et Electronique. With seven international editions it was also the world's largest. Its Indian edition is now the worlds' largest and most other now exist (under different names. See: https://en.wikipedia.org/ wiki/Electronics_Today_International

Collyn subsequently founded and published over 20 other publications in electronics, computing, music, and telecommunications - including Australian Communications. From 1982-1990 he was technology editor of The Bulletin and also Australian Business. In 1986, Collyn wrote the Australian Federal Government's Guide to Information Technology, and also the NSW Government's textbook for its electronics practique syllabus.

In 2018, Collyn entered into a business partnership with Daniel Weinstein - who is responsible for the major effort of converting all current and future books into eBook format and running our websites.

Daniel Weinstein has travelled through many parts of Australia with his wife Eve, various camper trailers, solar panels and copies of Collyn's books to fall back on. He's had a number of careers: professional puppeteer, owner of a graphic design/advertising company, software developer and website designer to name just a few. All of his previous careers (with the possible exception of puppeteer) have contributed to his latest career, converting this and all of Collyn's and Eve's books to an eBook and Print-on-Demand format to make them more available and accessible to a wider audience.

Acknowledgement

The author acknowledges the assistance provided by many companies in allowing their illustrations to be used here, and also advice provided by Redarc in connection with battery charging.

I especially thank Laurie Hoffman for her invaluable advice in the mobile telecommunications area. In keeping with the spirit of this book, it was originally done whilst she was camping alongside the Murray River in her hi-tech motorhome.

Table of Contents

Preface	1
Chapter1	3
Solar realities	3
Where can solar energy be used?	3
How is available solar measured?	3
Solar anomalies and limitations	6
Space and weight limitations	6
Converter electrical systems	7
Cabins	7
Battery capacity	8
Cooking and heating	9
Energy-efficient appliances	9
Water pumping	9
Washing machines/dishwashers	10
Television	10
Computers	11
Lighting	11
Air conditioning	11
What voltage?	11
Mains-voltage via an inverter	12
Costs	13
Avoid cheap products	13
Chapter2	14
Electrical self-sufficiency	14
Vehicle alternator charging	14
The self-sufficient approach	16
Jump starting	17
Future for alternator charging	17
Chapter3	18
Solar modules	18
Monocrystalline modules	19
Polycrystalline modules	19
Amorphous modules	19
Solar cell development	20
What solar modules really produce	20
Shadow resistance	22
Module placement	22
Portable solar modules	23
Buying solar generally	24
Grid-connect solar modules for cabins and RVs	24
MPPT regulators	25
Solar module trends	25
Chapter4	26
Solar regulators and monitors	26
MPPT regulators	27
Buying a solar regulator	28
Energy monitors - knowing the state of charge	28
How energy monitoring works	29
Chapter5	31
Batteries and battery charging (general)	31
Specialised batteries	32
Lithium-iron (LiFePO4) batteries	32
High current loads - caution	33
How batteries are charged	33
Typical battery charging sequence	34
Equalisation	35
Battery Management Systems	36
Chapter6	38
Batteries and battery charging (via alternators)	38
Temperature controlled alternators	38
Variable voltage alternators	39
Identifying alternator type	40
Battery-to-battery DC chargers	40
Charging lithium-iron batteries	41
Choosing a battery-to-battery charger	41
Adding solar	42
Need for buying caution	43
Chapter7	44
Generators and fuel-cells	44
Safeguarding electrics	45
Charging from an inverter generator	45
How noisy?	46
Diesel/LPG generators	46
Fuel-cells	47
Chapter8	48
Inverters	48
Inverter types	48
Transformer- based inverters	48
Adequate installation essential	49
Switch-mode inverters.	50
Paralleling inverters	50
Wired in - or freestanding	51
Automatic load sensing - 'phantom loads'	51
Safety - a buying consideration	53
Chapter9	54
Refrigerators	54
Top or door opening?	54
Electric-only	55
Eutectic fridges	56
Three-way fridges	57
T-rating	58
The choice	59

Correct installation	60
Chapter 10	61
Lighting	61
Switching	61
Halogen	61
Fluorescent lights	62
Light emitting diodes	62
Colour temperature	63
Light output	64
Lumens	64
Lux	65
Light fittings for RVs	65
The choice	66
Chapter 11	67
Water and pumping	67
Water pumps for RVs	67
Pipe resistance	68
Pressure tanks	68
Constant pressure pumps	70
Chapter 12	71
Computers and TV	71
Television	71
Computers	71
Housing equipment	72
Chapter 13	73
Communications	73
Yagi Antennas	74
Charging the phone	74
Satellite phone	75
High Frequency radio	76
Personal locator beacon (PLB)	77
CB radio	77
Useful Telstra links:	78
Chapter 14	79
Scaling the system	79
Table 1 - Typical consumption - in watts	80
Table 2.14	80
Two solar approaches	81
How many modules?	82
Supplementing the battery	83
Solar output - modules flat	85
Solar output - modules angled	85
Supplementing solar energy	87
Winter use	87
Electrical self-sufficiency	87
Chapter 15	89
Example systems	89
Watts the matter	89
Example 1. Small cabin	90
Example 2. Campervan (40 litre electric fridge)	91
Example 3: Small motorhome (120 litre electric fridge)	92
Example 4. Average caravan/motorhome	93
Example 5: Large caravan	95
Example 6: Coach conversion (old 500 litre fridge)	96
Example 7. Big fifth-wheel caravan - (low energy)	98
Chapter 16	100
Extra-low voltage wiring	100
The Victorian exception	100
Safety	101
Absolutes	102
Wire tables	104
Auto cable - a very real trap	105
Current rating	106
Specifying cables	106
ISO Standard re voltage drop	106
12/24 volts - the choice	107
Fixing inadequate wiring	108
Which cable is which - colour coding	108
Making connections	109
Joining cables	110
Fuses	111
Circuit breakers	112
Master switches	113
Switches	113
Twelve volt plugs and sockets	114
Chapter 17	115
Low voltage wiring	115
Relocatable premises	116
Permanent connection exceptions	117
Vital exceptions	117
Power into the vehicle (socket outlets)	118
Supply cables	118
10-15 amp adaptors	119
Protecting against 'dirty' power	119
Vehicle inlet socket	120
Polarity and double pole switches	120
Polarity testing	121
Inverters	122
Change-over switches	123
Cabling	123
Separation of Low and Extra-low voltage wiring	123
Wiring protection	123
Kitchens and bathrooms	124
Certification	124
Updating installations	124
Chapter 18	125
Installing batteries	125
Parallel connection	125

Series connection	126
Battery location and care	126
Battery installation and ventilation	127
Battery and cable protection	127
Battery cables and connections	128
Power posts	129
Sizing main battery cables	130
Safety precautions	131
Chapter19	132
Installing solar modules	132
Placing modules	133
Connecting loose modules	134
Roof mounting	134
Trailers/fifth-wheeler caravans	134
Physical mountings (RVs)	135
Cabins	136
Interconnection for higher voltage/current	137
Solar module cable sizing	138
Series diodes	140
Earthing	140
Chapter20	142
Installing the solar regulator	142
Load measurement	144
Current shunts	145
Positive versus negative (shunt) sensing	145
Chapter21	147
Installing the fridge	147
Example 1	148
Example 2	148
Chapter22	149
Installing an inverter	149
Connection sequence	149
Chapter23	151
Installing water systems	151
Pressure tanks	152
Chapter24	153
Installing a voltage sensing relay	153
Variable voltage alternator issues	154
Chapter25	155
Electrical converters	155
Converter associated wiring issues	156
Chapter26	157
Compliance issues	157
Electrical non-compliance	158
Chapter27	160
Fixing problems (general)	160
Chapter28	163
Living with solar	163
Typical indications of a good system	163

with lead acid or AGM batteries	
Typical indications of a system with lithium batteries	165
Solar capacity versus battery capacity	165
Training visitors	165
Chapter29	167
Walking the walk	167
VW Kombi	168
OKA	169
Nissan Patrol and TVan	169
Home and property	170
Swimming pool	171
Water pumping	172
Church Point	172
Chapter30	174
Electricity - simply explained	174
What is electricity?	174
General	175
Electrical units and terms	175
Chapter31	177
Making contact	177
Useful sites	177
This book's publisher	178
Acknowledgement	180